The Life Science Innovation Roadmap

The Life Science Innovation Roadmap

GETTING YOUR IDEAS TO MARKET

Arlen D. Meyers, MD, MBA and Courtney Price, PhD

LOGOS PRESS

The Life Science Innovation Roadmap

Arlen D. Meyers, MD, MBA and Courtney Price, PhD

Published in The United States of America
by
Logos Press®, Washington, DC
www.Logos-Press.com
info@Logos-Press.com

10 9 8 7 6 5 4 3 2 1

ISBN-13
Hardcover: 978-1-934899-26-7
Softcover: 978-1-934899-27-4

Library of Congress Cataloging-in-Publication Data

Meyers, Arlen D.
The life science innovation roadmap / Arlen D. Meyers and Courtney Price.
p. cm.
Includes bibliographical references and index.
ISBN 978-1-934899-26-7 (hbk.) -- ISBN 978-1-934899-27-4 (perfect bound)
1. Biotechnology industries. 2. Life sciences--Economic aspects. 3. Life sciences--Research--Economic aspects. 4. Entrepreneurship. 5. New products. I. Price, Courtney H. II. Title.
HD9999.B442M49 2012
338.4'7570973--dc23

2012014728

Contents

Preface

THE AUTHORS WROTE the *Life Science Innovation Roadmap: Getting Your Ideas to Market* because of our challenging times for the US bioinnovation and medical-industrial relationships. Skyrocketing US national debt, a global recession, high unemployment, geopolitical upheaval in the Middle East, a continuing dependence on fossil fuels, a distrust of government, street demonstrations against corporate greed and income inequality, the deteriorating status of US public education, and the uncertainty of US healthcare and regulatory reform has forced investors to the sidelines and threatened our "biomedical innovation pipeline." Our response to these threats will determine whether the US maintains its dominance in biomedical innovation or continues to yield that position to another country.

One part of the formula of reigniting the commercialization of life science ideas is educating a cadre of bioentrepreneurs who have the knowledge, skills, abilities, experience, and connections to get their ideas to market. Armed with primarily only a technical background in science, technology, engineering, healthcare, and math, life science innovators need entrepreneurial education and training that specifically addresses how to get a drug, device, diagnostic, healthcare IT, or service delivery model or platform to the market.

With that in mind, this book is intended for scientists, engineers, physicians, business people, and service providers at all levels who are interested in and support the process of life science commercialization. We wrote this book to provide our readers with a hands-on template for success based on our years of experience practicing and teaching bioentrepreneurship and consulting to clients, students, economic development agencies, and faculty around the world.

Section I: Bioscience Environmental Assessment includes Chapters 1 and 2 which lays the foundation for life science entrepreneurship. It provides an overview of the changing US and global healthcare landscape, the entrepreneurial opportunities they represent, and a synopsis of entrepreneurial fundamentals and the entrepreneurial mindset.

Section II: Planning & Strategy includes Chapters 3-9 which first introduces the Life Science Innovation Roadmap™ and then outlines strategies and helpful hints about how to successfully navigate it. This section also addresses how to conceptualize, protect, validate, and test your ideas reflecting the best and leading innovation practices. Lastly, this section addresses life science regulatory and reimbursement issues and how to build a strong business model.

Finally Section III: Execution & Implementation includes Chapters 10-15 which provide practical advice and recommendations about assembling your team, writing your plan(s), selling your ideas to stakeholders, financing and exit options for a new venture from the early stages to maturity. It also covers how to be a corporate innovator as an employee working for a healthcare organization.

Each of the chapters is richly supplemented with tips, examples and resources that represent core ideas and key principles including lessons learned not only from successful ventures but failures as well.

There are three main reasons why doctors, healthcare professionals, scientists, and engineers should embrace bioentrepreneurship and acquire the knowledge, skills, and abilities to practice it and participate in commercialization initiatives.

First, bioscience inventions do no one any good until they get into the marketplace. The bioscience innovation pipeline is broken; despite billions of dollars poured into basic research, relatively few true game-changers have emerged from the bottom of the funnel.

Second, you can make money, sometimes lots of it, by commercializing your idea as well as having a sense of self-fulfillment and accomplishment. There are several ways to participate, including working within your organization or with industry partners in various roles, licensing your ideas, or creating your own company. How you play is up to you.

Finally, bioentrepreneurship provides an outlet for doctors who want to use their entrepreneurial, intellectual, and problem-solving skills to make a difference and add another dimension to their lives and the lives of their patients. They can move from taking care of patients one at a time to creating a drug or device company that makes a cure for thousands and employs hundreds.

If you find joy and challenge in learning new things, are undaunted by the complexities of commercializing bioscience inventions, and don't mind working long hours you'll do well as a bioscience entrepreneur. We hope this book helps you navigate the pathway to life science commercial success.

Arlen D. Meyers, MD, MBA
Arlen.meyers@ucdenver.edu

Courtney Price, PhD
cprice@venturequestltd.com

Denver, Colorado
2012

Acknowledgements

THE AUTHORS WOULD like to offer a special thanks and recognition to Jill Hansen, Director of Business Development at VentureQuest, who we've worked together with for many years for her valuable suggestions and feedback vetting many of our book ideas while ensuring this project stayed on track.

We would like to thank our spouses, Kathleen and Gordy, for their support and encouragement. In addition we would like to thank the many bioentrepreneurs, innovators, and thought leaders that we've worked with over the years who shared their stories, insights, knowledge, and learning about the business development process. Lastly, we would also like to acknowledge the support and contributions of our global colleagues who are researching, teaching, practicing and advancing the art of bioentrepreneurship. The future of healthcare depends on them.

Section I
Bioscience Environmental Assessment

Chapter 1

Life Science Entrepreneurship and the Changing Healthcare Landscape

Life science entrepreneurship is the process of creating customer-defined value by researching, developing, and commercializing biomedical ideas, discoveries, and inventions. There are many new exciting business opportunities for innovators to develop and commercialize their new products and services. However, the commercialization processes is risky, expensive and time consuming. To be successful, bioentrepreneurs whether healthcare professionals, scientists, engineers, investors or service providers, need to work as a team and or with their organizations to overcome the multiple hurdles taking their ideas to the market and patients.

The process is neither linear nor predictable and outcomes are never guaranteed. In addition, because of global macroeconomic conditions, investors are unwilling to gamble on unproven technologies in a more hostile regulatory and legal environment. Consequently, commercializing bioscience discoveries is becoming more and more difficult. However, innovators still thrive.

Where are some of these exciting business opportunities for bioentrepreneurs? An initial understanding of the changes happening in US healthcare is the first step in identifying potential market opportunities.

THE US HEALTHCARE LANDSCAPE

Recent health care reform in the United States has been enacted via two bills: the Patient Protection and Affordable Care Act which became law on March 23, 2010 and was shortly thereafter amended by the Health Care and Education Reconciliation Act of 2010 (H.R. 4872) (which became law on March 30). In addition, the American Recovery and Reinvestment Act of 2009 (the stimulus bill) and the Health Information Technology for Economic and Clinical Health (HITECH) Act, enacted as part of the American Recovery and Reinvestment Act of 2009, was signed into law on February 17, 2009, to promote the adoption and meaningful use of health information technology. These bills, and others like it, ushered in the next phase of US healthcare reform. Reform efforts are designed to improve access, lessen the number of uninsured, lower costs, improve population health and improve outcomes. While the results will not be known for several years, the changing landscape has created opportunities and threats for those interested in biomedical innovation and healthcare.

SHRINKING EMPLOYER BASED HEALTHCARE COVERAGE

Between 2000 and 2009, the percentage of the population under 65 years of age with private health insurance obtained through the workplace declined from 67 percent to 58 percent.[1]

THE CONTINUING RISE IN HEALTHCARE SPENDING

Healthcare spending in 2009 grew at the slowest rate in five decades, rising 4 percent to $2.5 trillion, while still outpacing overall growth in the gross domestic product. Excessive costs without improvements in outcomes have driven major healthcare reform proposals. In 2014, 15 million new patients will become eligible for insurance coverage.

---

1 "Health, United States, 2010," *National Center for Health Statistics/Centers for Disease Control and Prevention*, February 2011, http://www.cdc.gov/nchs/data/hus/hus10.pdf

HEALTH INSURANCE REFORM AND CHANGES TO THE REIMBURSEMENT MODEL

The traditional fee-for-service model of care delivery is under attack. Critics claim that the present system encourages overuse of costly services, and that, in some instances, are unnecessary, redundant or harmful. Under prodding from the Centers for Medicare and Medicaid Services (CMS), the payment models are being changed to encourage payment for favorable outcomes and care that adheres to best evidence and cost-effective practice.

For example, an Accountable Care Organization (ACO) is a health system model with the ability to provide, and manage with patients, the continuum of care across different institutional settings, including at least ambulatory (outpatient) and inpatient hospital care and possibly post-acute care. Further, ACOs have the capability of planning budgets and resources and are of sufficient size to support comprehensive, valid, and reliable performance measurement. The ACO model is one of the latest designs for managing healthcare costs and especially Medicare costs, and is gaining traction among policymakers desperate to control costs and boost quality in healthcare.

The healthcare reform law placed new requirements on health insurers.

- Beginning in 2011, insurance companies must spend at least 80 percent of premium dollars on medical care and quality improvement activities. This figure, known as the medical loss ratio, is 85 percent for insurers in the large-group market.
- Insurers that fail to meet the ratio must pay rebates to consumers by August 1, 2012. The government estimates that up to nine million Americans could be eligible for rebates worth up to $1.4 billion.
- "Mini-med" health plans, which have low annual coverage caps, get an exception that will make it easier for them to meet the required ratios.

The results will be changes in how and how much doctors, medical device and pharmaceutical manufacturers will get paid for their products and services.

Examples of how this will impact providers are failure of insurance companies to pay for implants used in ambulatory surgery settings, modification of payment rules for oncologists administering chemotherapeutic agents in their offices and restrictions on payment for imaging studies done inappropriately.

FRAGMENTED CARE TO THE PATIENT CENTERED MEDICAL HOME

The Patient Centered Medical Home (PCMH) is an approach to providing comprehensive primary care for children, youth and adults. The PCMH is a health care setting that facilitates partnerships between individual patients, and their personal physicians, and when appropriate, the patient's family.

DOWNSIZING /RIGHTSIZING THE HEALTHCARE WORKFORCE

Reimbursement reform, the economic recession and rising unemployment has resulted in more patients who are uninsured or underinsured, patients and employers paying higher copays and deductibles, and, as a result, lower utilization of some healthcare services. As a result, hospital admissions in some locations have dropped, forcing hospital officials to cut costs and reduce the number of staff.

Coincidentally, several experts are predicting nursing and physician manpower shortages. For example, researchers predict a shortfall of RNs developing around 2018 and growing to about 260,000 by 2025 and, by 2025, the nation will be short 35,000 to 44,000 adult-care generalists practicing family medicine and general internal medicine.[2] For example, in Colorado, 500,000 will added to the health insured rolls in 2013, requiring about 150 new primary care doctors.

---

2 http://healthaffairs.org/blog/2010/06/14/health-care-workforce-nurse-and-physician-shortages/

ELECTRONIC MEDICAL RECORDS, HEALTHCARE INFORMATION EXCHANGES AND DATA ANALYTICS/BUSINESS INTELLIGENCE

Perhaps no other area has undergone as much change as those efforts to create integrated systems to document episodes of care using electronic medical records. In an effort to create a national healthcare information highway, the government has mandated use of electronic medical records and created incentives and penalties to drive its adoption. In addition, integrated healthcare exchanges (HIE's) have been created to interconnect systems in regions throughout the United States. Data analytical and business intelligence tools are designed to mine the data in the networks and, by integrating financial data with clinical data, identifying outcomes and creating management dashboards.

ACUTE TO PREVENTIVE CARE

There is increasing emphasis on preventive care, wellness and disease management. For example, Medicare has approved payment for 3 visits a month for 3 months to give advice and care to obese patients. New companies and models have emerged to take care of patients away from the hospital whenever possible. In addition, eliminating readmissions to the hospital, particularly for patients with respiratory diseases and heart failure, is a priority.

MOBILE HEALTH

Mobile health refers to transferring or accessing healthcare information using mobile communication devices like smart phones, laptops, iPads or similar devices. According to Kalorama Information, the market for mobile health apps reached $84.1 million in 2010.[3] The industry continues to evolve and mature.

For example, WellDoc's DiabetesManager System for care providers and adult patients with type 2 diabetes received FDA clearance during the quarter. Five years in the making, WellDoc's offering is a mobile health system built on automated clinical coaching and behavioral algorithms driven by real-time patient data, according to the company. WellDoc's software aims to enable care providers to

3 "The worldwide market for mobile medical apps report" *Kalorama Information*. Pub ID: KLI2831262

extend their care beyond traditional office visits by utilizing mobile phones and the Internet. The Diabetes Management system specifically features a medication adherence program and secure capture, storage, and real-time transmission of blood glucose data. WellDoc's service doesn't just help users track their biometrics, the company also runs the data through its own proprietary Automated Expert Analytics System, which identifies trends and delivers relevant educational and behavioral patient coaching based on the data.

INCREASING PRIVATE INVESTMENT IN THE HEALTHCARE INDUSTRY

Federal healthcare reform legislation has attracted the attention of private investors in the healthcare industry, separate from drugs, devices and diagnostics. According to a survey in 2010, 11 percent of private equity executives planned to invest in healthcare service facilities, up from 4.8 percent only six months earlier.[4]

THE GROWTH OF EMPLOYED PHYSICIANS

Driven by the need to integrate and coordinate care, hospitals are increasingly employing physicians and buying existing physician practices and networks. Over 65 percent of physicians in the US are now employed by someone. This change in hiring and practice patterns will create new threats and opportunities for health care system employers.

THE GROWTH OF TELEMEDICINE AND TELEHEALTH

According to the American Telemedicine Association, telemedicine is the use of medical information exchanged from one site to another via electronic communications to improve patients' health status. Closely associated with telemedicine is the term "telehealth," which is often used to encompass a broader definition of remote healthcare that does not always involve clinical services.

Videoconferencing, transmission of still images, e-health including patient portals, remote monitoring of vital signs, continuing medical education and nursing call centers are all considered part of

4 Shinkman, R. "M&A roundup: Private equity charges into hospital market" August 29, 2011. http://www.fiercehealthfinance.com/special-reports/ma-roundup-private-equity-charges-hospital-market

telemedicine and telehealth.

Telemedicine is not a separate medical specialty. Products and services related to telemedicine are often part of a larger investment by health care institutions in either information technology or the delivery of clinical care. Even in the reimbursement fee structure, there is usually no distinction made between services provided on site and those provided through telemedicine and often no separate coding required for billing of remote services.

Telemedicine encompasses different types of programs and services provided for the patient. Each component involves different providers and consumers.

- Specialist referral services typically involves of a specialist assisting a general practitioner in rendering a diagnosis. This may involve a patient "seeing" a specialist over a live, remote consult or the transmission of diagnostic images and/or video along with patient data to a specialist for viewing later. Recent surveys have shown a rapid increase in the number of specialty and subspecialty areas that have successfully used telemedicine. Radiology continues to make the greatest use of telemedicine with thousands of images "read" by remote providers each year. Other major specialty areas include: dermatology, ophthalmology, mental health, cardiology and pathology. According to reports and studies, almost 50 different medical subspecialties have successfully used telemedicine.
- Patient consultations using telecommunications to provide medical data, which may include audio, still or live images, between a patient and a health professional for use in rendering a diagnosis and treatment plan. This might originate from a remote clinic to a physician's office using a direct transmission link or may include communicating over the Web.
- Remote patient monitoring uses devices to remotely collect and send data to a monitoring station for

interpretation. Such "home telehealth" applications might include a specific vital sign, such as blood glucose or heart ECG or a variety of indicators for homebound patients. Such services can be used to supplement the use of visiting nurses.

- Medical education provides continuing medical education credits for health professionals and special medical education seminars for targeted groups in remote locations.
- Consumer medical and health information includes the use of the Internet for consumers to obtain specialized health information and on-line discussion groups to provide peer-to-peer support.

INNOVATION MANAGEMENT SYSTEMS

Healthcare innovation management systems are designed to identify, process, and internally and externally develop ideas, inventions or discoveries that have potential commercial value. Most large research intensive university health sciences centers and universities have mature technology transfer offices to manage this function. However, most large community-based hospitals and hospital systems do not. Since more doctors, nurses, pharmacists and other biomedical knowledge workers are employees of hospital systems, when properly structured, the intellectual property of these workers is owned by the employer, with the potential to monetize these ideas.

THE GROWTH OF NON-US EMERGING MARKETS

The pharmaceutical market in China is beginning to gain serious momentum as it is predicted to grow to more than $50 billion next year, making it the world's third-largest pharmaceutical market according to the annual forecast by IMS Health. In addition, the manufacture of biosimilars and generic drugs has migrated to India and Israel, making global biopharmaceutical manufacturing the norm.

CHANGES AT THE FDA AND INCREASING REGULATORY OVERSIGHT

Increasing FDA oversight for biosimilars and new 510(k) regulations are two examples of increasing regulatory oversight that has recently occurred. In addition, the process for approval of devices in the US takes almost two years longer than in Europe. Consequently, investors have been fleeing from the biotech sector and inventors are going overseas to research, develop and commercialize their inventions in an effort to get their products to market faster.

CHANGING PHYSICIAN ROLES: HOSPITALISTS[5], INTENSIVISTS[6] AND CONCIERGE MEDICINE

Founded by physicians in 1995, IPC is the nation's leading national physician group practice focused on the delivery of hospital medicine and related facility-based services. IPC's providers, support staff, operations, and technology are all focused on driving the health care delivery system towards better, more efficient and high quality inpatient care. Today the company manages over 150 practice groups, providing care in over 500 medical facilities in 21 states.

IPC has taken a leadership role in the emergence of this new medical specialty and today has over 1400 affiliated providers in the US. IPC works closely with approximately 30,000 referring primary care physicians and specialists and more than 1,000 health plans, to manage more than $1 billion in healthcare expenditures annually.

THE AGING POPULATION

The US population is aging and accounting for increasing expenditures. In 2008, adults 85 and older accounted for only 1.8 percent of the population but 8 percent of all hospital discharges; 75- to 84-year-olds comprised 4.3 percent of the population but accounted for 13.8 percent of all hospital discharges.[7]

5 physicians who only care for patients in a inpatient setting

6 specialists in critical care who take care of patients in intensive care units

7 "Hospital utilization among oldest adults, 2008," *Healthcare Cost and Utilization Project*

HEALTHCARE GLOBALIZATION AND MEDICAL TRAVEL

Leaving home for care, whether it is domestically or internationally, has grown to a multibillion global industry. Patients are getting dental, medical, wellness and alternative and complementary medical treatments because they are cheaper, of high quality and more readily accessible than in their home countries. In addition, price and quality data transparency is empowering patients with the information they need to make comparable choices and global telemedicine and healthcare information exchange makes overseas access easy.

Medvoy (www.medvoy.com), for example, is a physician to physician global referral and communications platform for doctors practicing international care. Careconnectix (www.careconnectix.com) is a global healthcare social network for patients and healthcare professionals around the world. Companies supporting the global care ecosystem will continue to appear.

THE EMERGENCE OF SOCIAL MEDIA IN HEALTHCARE

Healthcare organizations and physicians have embraced social media, like Facebook, LinkedIn, and YouTube to communicate with patients, help them manage disease, recruit new physicians and get intelligence about products and services. In addition, it is used to educate patients and collaborate with them not just in clinical care, but in research and development as well. Many questions remain about how social media should be used in healthcare and how to measure the return on investment. What is clear, however, is that the use of social media, tied to both mobile health and other online platforms will continue to evolve and represent big business opportunities.

PATIENT DRIVEN RESEARCH AND DEVELOPMENT

Biomedical research is, for the most part, designed and conducted by researchers and companies, recruiting patients to enter trials. However, the research agenda and control of information has started to shift to patient groups. Patientslikeme.com, for example, is a crowd sourcing site consisting of patients willing to share their experiences taking experimental drugs, dealing with unusual or rare diseases, or promoting awareness among other patients with incurable or untreatable diseases, like progressive neurological diseases. Several Internet sites attempt to raise money from private sources for biomedical research. For example, womensheart.org raises money to do research and raise awareness of women with heart disease.

SUMMARY

- The global healthcare and biomedical research and development landscape is changing quickly.
- All of these changes present biomedical and healthcare entrepreneurs opportunities to create new products, services, models and platforms.
- Patients are taking more control of funding and contributing to basic and clinical research using the Internet and social media.
- The Internet and social media continues to play a bigger and bigger role in healthcare marketing and delivery
- Funding for early stage ideas is harder. As a result of global macroeconomic conditions, changing business models, and increasing regulatory oversight, investors have “fled to quality” and are risk averse. Biomedical entrepreneurs need to be persistent and creative when it comes to raising money for new ventures.

Chapter 2:

Bioentrepreneurs and Bioentrepreneurship

Does this scenario sound familiar? In the operating room, you use an instrument that is unwieldy or does not completely get the job done. At the bedside, you waste a lot of time changing dressings because all the wound care materials are not in one convenient place. Or, while seeing patients in your office, the information you need is not right at your fingertips. Then you ask yourself, "Why doesn't someone invent something to fix that?"

International scientists, bioengineers, healthcare IT experts, and healthcare professionals in all specialties and areas of interest are accepting that challenge. Practitioners are in a unique position to identify unmet healthcare needs and work with basic science, engineering, IT, and business partners to develop and commercialize solutions. This results in the creation of new drugs, devices, diagnostics, health information technologies, healthcare delivery innovations, and new healthcare business models.

Entrepreneurship has many definitions. Bob Reiss, a successful entrepreneur and author of *Low-Risk, High-Reward: Starting and Growing Your Small Business with Minimal Risk*, is one of several who agree that:

> Entrepreneurship is the recognition and pursuit of opportunity without regard to the resources you currently

control, with confidence that you can succeed, with the flexibility to change course as necessary, and with the will to rebound from setbacks.

We define bioentrepreneurship as the process of creating customer- defined value by researching, developing, and commercializing biomedical and healthcare innovations. The definition includes three key elements: process, value and innovation.

- Bioentrepreneurship is a process

  Similar to all entrepreneurs, bioentrepreneurs mostly act. Indeed, they plan and monitor their activities but the process is proactive and hands on. Innovators are constantly experimenting and rapidly adapting to changing macroeconomic stresses, market shifts, and technological advancements. Seeing opportunity and exploiting it, experimenting, rapidly changing course, and bootstrapping are core bioentrepreneurial competencies.

- Bioentrepreneurs create customer-defined value

  To be successful, the innovative product or service must have tangible and intangible benefits in the mind of the buyer that far exceed the tangible and intangible costs. Moreover, the customer has to define the offering's value, not the bioentrepreneur. Customers vote with their wallets.

- Bioentrepreneurs innovate

  Bioentrepreneurs innovate by creating things that are new, better, smarter, faster, cheaper, and/or are more attractive than the status quo, the competition, or the alternatives. In some instances, the difference is incremental and minimal. In others, it is disruptive and entirely novel.

BIOENTREPRENEURSHIP IS NOT PRACTICE MANAGEMENT

Some healthcare professionals confuse practice management with bioentrepreneurship. There is a big difference between the two. The goal of practice management is to run a medical practice so that it is efficient, effective, scalable, profitable and sustainable. Practice management is the operations side of the business, just like any other industry. The goal of bioentrepreneurship, on the other hand, is to create value in the marketplace by exploiting biomedical product or service innovation. Practice management is part of bioentrepreneurship, but a means towards an end, not the end itself. Once an entity is created, it needs to run profitably to be sustainable and grow. Practice management is, by definition, a management exercise. Bioentrepreneurship is an innovation, creativity, and leadership activity. Each of these activities requires very different knowledge, skills and abilities. Few people, not just healthcare professionals, are equipped to do both. Moreover, providing doctors, dentists, nurses, or pharmacists practice management education or all the MBA courses in the world will not necessarily result in more innovation.

In addition, entrepreneurship is but one element of innovation. Furthermore, innovation is only one element of competitiveness, either locally or globally. Competiveness requires several elements in addition to innovation, for example, intellectual property protection, laws that protect private property, and a stimulating macroeconomic environment that rewards risk and provides capital. Bioentrepreneurship is about creating new products and services. Practice management is about running the business.

MARKET PERCEIVERS AND TECHNOPRENEURS

Life science entrepreneurs are either technopreneurs (technical professionals, life scientists, healthcare professionals, or engineers) or market perceivers (investors, business professionals, or service providers.) There are significant differences in how the two approach the process or life science commercialization and how they see the world.

Technopreneurs are primarily solution driven. They typically invent or discover something and then search for an application or market. They adopt a technology push approach. Market perceivers, on the other hand, start with uncovering or sensing a market need, and then seek to develop a product or service that will satisfy customers and solve their problems. They use a market pull model. Whether a technopreneur or market perceiver, there are four basic models of bioentrepreneurs that are differentiated by their size, scope, scale and goals.[1]

THE FOUR TYPES OF ENTREPRENEURSHIP

Not all life science entrepreneurs have the same goals, exit strategies or stakeholders. Steven Blank developed a model describing four types of entrepreneurs.[2] His model applies to bioentrepreneurs, whether they are technopreneurs or market perceivers, as well.

1. **Private or independent practice**

   Most research shows that about 40 percent of US doctors and most dentists are independent medical practitioners who own their own businesses. However, the situation for medical doctors is changing as more and more doctors become employees of healthcare organizations. The reasons for the increasing numbers of employed physicians are many and include practice management hassles, student debt, work-life attitudes, generational attitudes about work, and differences in how the genders practice.

   Like other non-medical small businesses, the primary goal of the business is to generate enough profit to satisfy the needs of the owners. The scope and scale of the practice is typically limited by the ability of any one doctor to see a certain number of patients per day. In addition, the exit often involves the sale or transfer of the practice once the owner retires or dies.

---

1 Shreefal, M. "Paths to entrepreneurship in the life sciences" October 26, 2004. http://www.nature.com/bioent/building/entre/102004/full/bioent831.html

2 http://steveblank.com/2011/02/08/startup-america-dead-on-arrival/

While some private, independent practices generate substantial revenues, they are less than large scalable companies, and the practices are typically funded by retained earnings. The intent of the owner is not to scale the practice to as large as possible, but to maximize revenues given finite time and resources.

2. **Scalable Startup Entrepreneurship**

   Unlike private practices, scalable healthcare and biomedical startup companies attempt to commercialize drugs, devices, diagnostics, healthcare IT products, new business models, and expanded service platforms. These bioentrepreneurs start a company with a vision that could change the world. Their job is to search for a repeatable and scalable business model. The goal of the scalable startup entrepreneurship is to create wealth for shareholders.

3. **Intrapreneurship**

   Large companies, like multihospital healthcare systems or chains of healthcare service products, have finite life cycles. Most grow through *sustaining innovation*, offering new products that are variants around their core products. As more healthcare knowledge workers become employees of hospital systems and healthcare delivery organizations, they become "intrapreneurs", meaning they mine new ideas and underutilized value within the organization. While inventions and discoveries can occasionally result in licensing revenue or spin out opportunities for the employer, more typically intrapreneurs contribute to or devise process improvements that eliminate waste and improve outcomes or service line extensions that are designed to offer new diagnostic or therapeutic innovations to patients. Changes in patient tastes, new technologies, legislation, new competitors, and so on can create pressure for more *disruptive innovations, that is* innovations that create entirely new, unheard of products/services sold into new markets or novel, unprecedented ways to sell existing products/services. Endoscopic surgery is an example of a disruptive innovation. It completely transformed the way surgeons performed historically invasive procedures. Device manufacturers capitalized on new products to improve endoscopic techniques, and hospitals had to readjust many of their existing systems, protocols, and medical equipment to fulfill patients' demand of the minimally invasive endoscopic approach.

4. **Social Entrepreneurship**

   Social entrepreneurs are innovators who focus on creating products and services that solve social needs and problems like eliminating smallpox or malaria or creating non-profit organizations or educational programs. Unlike scalable startups, the social entrepreneur's goal is to make the world a better place, not to capture market share, or to create to wealth for the company. They may be nonprofit, for-profit, or a hybrid.

Using this model, most practitioners are private practice entrepreneurs. Some have a need to perfect their practice management skills. Some physician entrepreneurs create new products, services, and/or platforms and seek to scale the business for an eventual profitable exit. Some practitioners, like those who work for large healthcare organizations, seek to improve processes or create new delivery models to drive growth. Finally, some doctors are social entrepreneurs, seeking new ways to solve social needs and problems.

UNIQUE CHARACTERISTICS OF THE HEALTHCARE MARKETPLACE AND INDUSTRIES

While the principles and practice of entrepreneurship can be generally applied to most industries, biomedical science commercialization has some unique challenges. They are:

- A highly regulated environment that is increasingly complex and expensive;
- A need to protect intellectual property to a greater degree than other industries;
- A convoluted system of third party reimbursement for services, drugs, and devices;
- Multiple stakeholders including patients, payers, providers, and industry;
- Rapidly shifting markets due to escalating costs, quality problems and reduced access to care;
- A capital intensive industry;
- A high failure rate for early stage companies due to the uncertain outcomes of pre-clinical and human clinical trials;
- Poor acquisition, interoperability, and integration of healthcare information;
- A globalized research, development, and commercialization ecosystem;
- Poor quality and price transparency;
- Difficulty comparing the effectiveness of one drug or device to another;
- Complex translational research and clinical trial environment; and
- Bottlenecks in the innovation and drug discovery process.

While these are all significant barriers to success, they are also big opportunities to innovate. Since change creates market gaps, savvy bioentrepreneurs and scientists fill them with new ideas, discoveries, and inventions.

IDEAS, INVENTIONS, AND INNOVATION

Bioentrepreneurs create life science innovations by exploiting limited resources like many other entrepreneurs. However, there are big differences among dreams, inventions, and innovation.

Dreams are ideas that pop into one's head and happen all the time. Leonardo Da Vinci was one of most famous dreamers or all time. He conceptualized helicopters, mechanical devices, and many other contraptions hundreds of years before they were actually built. He invented few of them.

Inventions, on the other hand, are ideas that are reduced to practice. In other words, thinking of a new design and creating it or taking existing services or processes and adding a new way to which then adds value to them.

Innovation is the process of taking inventions and discoveries and creating customer-defined value in the marketplace. Most of the time, the value is measured in money. In some instances, however (*e.g.* a creative way to increase vaccination rates in third-world populations); the metric is eliminating a disease like smallpox. Whether for-profit or not-for-profit, creating innovation is the difficult process of identifying a market niche, developing a product or service with a compelling value proposition (*i.e.* the promise the business makes to the customer) and implementing a plan so the invention makes a profit, sustains itself, or is subsidized. Merely building a device or creating a new healthcare service is worthless without a market willing to buy or use it. This is a critical concept that is frequently ignored by most life concept innovators.

There are three main reasons why doctors, scientists and engineers, whether they are in private practice, industry, or academic medicine, should embrace bioentrepreneurship and acquire the knowledge, skills, and abilities to practice it.

1. Bioscience inventions provide zero benefits until they get into the marketplace. However, the bioscience innovation pipeline is broken. Despite billions of dollars

poured into basic research, relatively few true game changers (*i.e.* disruptive technologies) have emerged from the bottom of the funnel.

2. Money can be made by commercializing ideas. There are several ways to participate, including working with industry in various roles, working with employers, licensing ideas, or creating new companies.
3. Bioentrepreneurship provides an outlet for doctors who want to use their entrepreneurial, intellectual and problem solving skills to make a difference and add another dimension to their lives and the lives of their patients. They can move from taking care of patients one at a time, to creating a new company that developed a cure for thousands of patients and creates jobs for hundreds of people.

ENTREPRENEURIAL KNOWLEDGE, SKILLS, AND ABILITIES

There is not a single entrepreneurial personality or makeup. Much research and experience indicates that almost any person can become an entrepreneur, regardless of their educational background, Myers-Briggs inventory category, Enneagram type, or personality profile. However, there are key behaviors that do seem to correlate with entrepreneurial success.

1. **Risk Profile**

   Entrepreneurs are not risk takers; they are risk managers. They carefully assess the returns based on a certain level of risk and attempt to maximize the yield. Of course, each entrepreneur differs in his/her level of risk taking. Some are risk averse, while others are willing to "bet the farm" on a big idea or market.

2. **Teamwork**

   Entrepreneurship is a team endeavor. Successful entrepreneurs recruit team members who are brighter, more talented, and more experienced than themselves. They know they cannot do everything successfully on their own.

3. **Leadership**

   The leader of the enterprise, the entrepreneur, must provide vision, direction, and inspiration for their team and company to be successful.

4. **Multitasking**

   Creating a new venture with limited resources requires much multitasking until revenues allow hiring people to fill organizational positions.

5. **Emotional intelligence**

   Successful entrepreneurs have highly developed self-awareness and people skills which they rely on to build alliances, secure trust, and safeguard trusting relationships. Entrepreneurs tend to be optimistic, independent, resourceful, and resist control and oversight.

6. **Passion**

   Entrepreneurship requires an inner drive to succeed and a single-minded dedication to the success of the opportunity. Passion is the emotional component of both the leadership team and the enterprise and goes beyond the strategy, tactics, and spreadsheets. Employees, investors, partners, and other potential stakeholders take measure of enthusiasm and zeal as a key indicator of how emotionally invested the entrepreneur is in the opportunity.

7. **Perseverance**

   The road to entrepreneurial success is paved with

failures, barriers, obstacles and impediments that will challenge an entrepreneur's problem-solving skills and persistence. Entrepreneurs have a bias toward action and tend to persevere until they find a successful solution. This constant experimentation allows the entrepreneur to quickly abandon opportunities that have fatal flaws but learn from their mistakes for the next time. They are curious, ask questions, and probe why something does or does not work or what could be improved.

8. **Excellent Networking Skills**

   Successful entrepreneurs are great networkers who like to build robust personal and professional networks of people who have different interests, knowledge, and connections than their own.

SUMMARY

- Bioentrepreneurship is the process of creating customer-defined value by researching, developing and commercializing life science innovation.
- Bioentrepreneurs are either market perceivers or technopreneurs
- Bioentrepreneurs tend to thrive within four different environments.
- Bioentrepreneurs tend to think, act, and feel in characteristic ways.

Section II
Planning and Strategy

Chapter 3:

The Life Science Innovation Roadmap™

INTRODUCING A NEW life science innovation to the marketplace is a journey that starts with the bioentrepreneur's vision. It is a rewarding and extremely challenging process, not for the "faint-of-heart." One of the major hurdles is the risk and uncertainty associated with developing new products/services for the marketplace. Another hurdle is ensuring that the product/service will carve out a sustainable competitive advantage. Then, there's the hurdle of determining if it can be sold at a profit, to whom, and who will be the first customers.

Launching a new life science idea in the 21st century can be a slow and sometimes dysfunctional process where you will move through a series of disconnected steps, tasks, and confusing hurdles. These hurdles can be time-consuming, frustrating, and frequently result in an abandonment of your idea. At minimum, such hurdles can impede an efficient and successful launch. Meticulous business planning, experimentation, validating assumptions, and just plain lucks can help inventors successfully overcome these challenges.

BUSINESS PLANNING

A little basic business planning goes a long way towards verifying if there is a real market for an invention. Business planning helps separate the winners from the losers. When bringing new technologies to the marketplace or starting any new venture, there are many failures. For example, according to the Small Business Association, only about 55 percent of new ventures survive four years.[1]

This is particularly applicable to the life science industry since it is highly competitive and frequently requires large capital investments. This research shows that the more planning and information gathering that can be done before the business opens, the greater the chance of long-term success. From 2004 through 2010, a study found that the overall success rate for drugs moving from early stage Phase I clinical trials to FDA approval is about one in ten, down from one in five or six seen in reports involving earlier years.[2]

Leading business schools and corporations tout business plans as a panacea to launching new business opportunities and showing investors and/or corporate decisions makers how successful an idea will be. Historically, business professionals and investors required some form of a business plan and maintained that preparing one is an essential exercise to articulate an idea's vision and plan its implementation. Investors proclaim that developing a strong business plan significantly increased the probability of obtaining outside capital. Executives usually required enterprise innovators to write a business plan before committing organizational resources to implement an idea. Academicians assert that writing a sound business plan will save an entrepreneur thousands or possibly millions of dollars that might otherwise be wasted on a flawed idea.

Due to this, all too frequently after the discovery of the new business opportunity, inventors hastily rush off to write a business plan which usually is flawed from the very beginning and a waste of precious time, money, and resources. Consequently, most business plans are written far too early in the business planning process and are often based on an ineffectively vetted opportunity. The plans are filled with invalidated assumptions drawn from research which may or may not be accurate and only paint part of the big picture. Inventors often waste time polishing elaborate business plans that include inflated financial projections, overly enthusiastic market research

1 B. Headd, (2003) "Redefining business success: Distinguishing between closure and failure," *Small Business Economics*, 21:51-61.

2 Industry survey: Clinical study budgeting practices and metrics. *ClearTrial*, August 2011.

data, and intricate details on day-to-day operations. Skeptics argue that business plans written too early often lack significant substance.

Many life science discoveries and inventions should have been dropped early during the planning process. Instead, the inventor develops a business plan much too early before the idea has been properly vetted. Oftentimes, an executive makes a gut business decision to move forward based on conjecture, or the executive and his/her colleagues have fallen in love with an invention. The steps of thoroughly and repeatedly evaluating the invention to prove that the business opportunity has strong potential are missing from this scenario.

A NEW APPROACH TO BUSINESS PLANNING

History has shown that developing business plans too early in the process is usually fraught with mistakes and risks and results in many failures. Different business planning processes have been developed, researched, and have evolved over time. The "older" business planning method, similar to the scenarios illustrated above, consist of discovering a new invention, writing a business plan, presenting the plan to decision makers, launching the new invention, and then trying to sell the new product or service. By contrast, the"newer" method first evaluates the opportunity's potential and then a feasibility plan is written to verify its commercial potential. During these early stages, inventors try to find market acceptance and have potential customers beta test the invention. The newer process validates the invention to determine if a business plan is warranted. Figures 1a and 1b below depict the older and the newer methods of developing business plans.

As these figures illustrate, business plans have not disappeared in the newer method. They are usually not developed until later in the planning cycle after the invention is thoroughly evaluated and the inventor tries to sell it to customers on a small scale. When business plans are written too early in the planning process, inventors skip the preliminary yet critical steps of assessing the invention's real commercial potential. In most cases, the feasibility plan filters out weak opportunities that have little chance of success. It will also help determine whether the invention should be licensed, sold, or

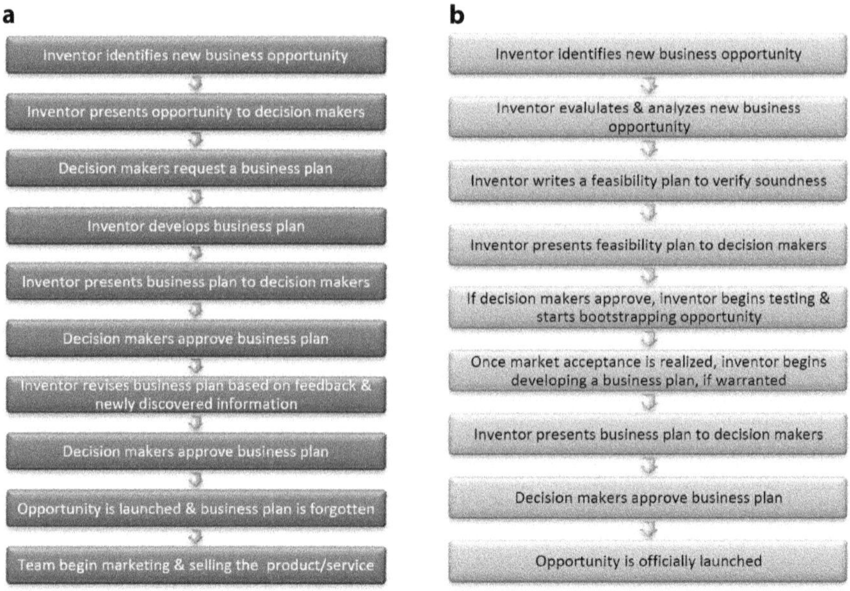

Figure 1: a) Typical older business planning method; b) Typical newer business planning method.

whether creating a joint venture is a better commercialization strategy. Skipping vital steps in the business planning process increases the chances of wasting time, money, and resources on an idea with little commercial potential.

Moreover, only a small percentage of inventions merit a formal business plan. However, there are some exceptions. Business plans should be written if a new venture will be created around the opportunity and/or if funding is being sought by investors. Likewise, banks typically require a business plan from a new venture when founders initially establish a banking relationship with them. Also, some governmental financing programs like the Small Business Administration (SBA) requires a business plan with funding applications.

When formal business plans are required, having developed a sound plan is important. Inventors should keep in mind that many of the projections and forecasts in the plan are often guesstimates. However, all projections should be based on sound budget and revenue forecasting assumptions that are comparable to other firms in the industry. Also, a business plan is a living document that changes over time which necessitates constant reexamination and updating. Likewise, once a new venture is launched, assumptions and projections may be somewhat realistic or inaccurate.

INTRODUCTION TO THE LIFE SCIENCE INNOVATION ROADMAP™

As a result of over 50 years of combined experience working with entrepreneurs, inventors, and high growth companies commercializing new products and services, we developed the Life Science Innovation Roadmap™ (LSIR) depicted in Figure 2. Use it to help guide you through the ups and downs of the business planning process.

As Figure 2 shows, the Life Science Innovation Roadmap™ has four distinct steps:

1. Idea Generation
2. Idea Assessment
3. Feasibility Analysis
4. Commercialization Path

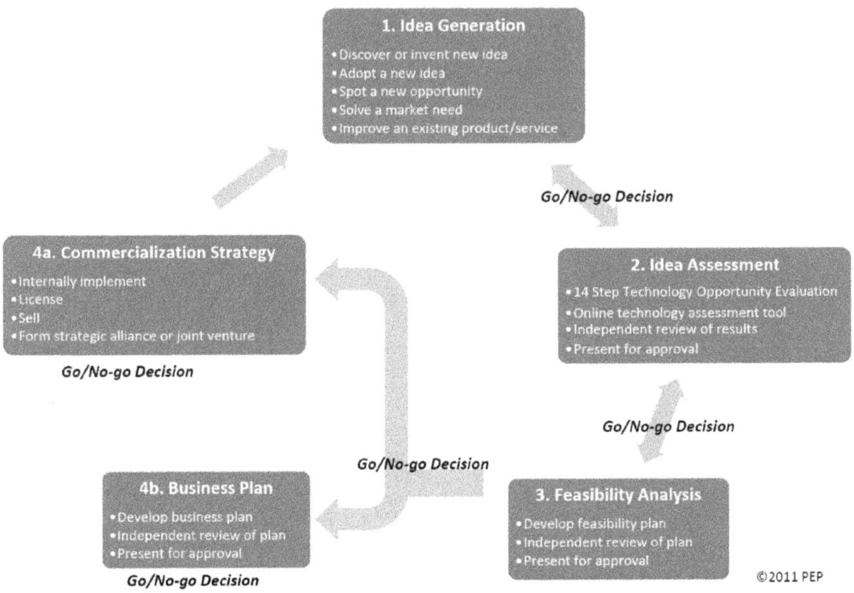

Figure 2: Life Science Innovation Roadmap

Innovation is a non-linear, two-way process where new technologies can skip a few steps or return to previous steps as it is diligently evaluated and new information is uncovered. The LSIR helps bioscience entrepreneurs establish critical milestones for an invention's successful commercialization. The Life Science Innovation Roadmap™ also helps innovators understand and master what it takes to commercialize a new product or service.

New technologies most often start out as ideas or discoveries progressing through the LSIR during which many go/no-go decisions are made. The decision makers could be a group of physicians, administrators, clinic directors, investors, lenders, or others who must provide approval to move forward. Regardless of whom the "go/no-go decision maker" is at each step in the Roadmap, innovators must be able to demonstrate to them the idea is valuable and worth further investment of time and money.

STEP 1

During Step 1, Idea Generation, new and exciting ideas are explored and identified. Oftentimes new ideas are discovered by finding new or better ways to solve a simple patient problem, identifying patients' needs or new market applications, and/ or fixing what is broken. Obtaining sufficient patient or customer knowledge and understanding is the key to discovering and designing new healthcare products and services.

Some health care providers schedule a brainstorming activity during their annual strategic planning event to generate new revenue-generating ideas. Other larger organizations might use an idea suggestion platform or some appoint an Innovation or Business Development Committee. See Chapter 4 for more information about generating new ideas. If the new idea is an invention, ensure that the discovery is properly documented in a lab notebook (see Chapter 7 for more information).

STEP 2

In Step 2, Idea Assessment, the innovation is initially evaluated to determine if it has market potential. Our 14 Step Technology Opportunity Evaluation (see Chapter 5) is an initial, quick assessment tool to determine if the idea has any commercial potential. This exercise helps innovators make the go/no-go decision to either more rigorously assess the innovation or search for other, more attractive ideas. If the innovation moves forward, more in-depth assessments should be completed. In addition, Chapter 6 includes some useful tips to help innovators articulate the commercial promise an idea holds.

Speaking with subject matter experts and other knowledgeable people (without giving away your intellectual property) can provide valuable insights about an innovation's technical feasibility and market potential. During this step, innovators must demonstrate that the innovation works as intended and has a defined market and potential customers. After completing this initial screening, a go/no-go decision is made whether to proceed or drop the innovation, shelve it, or develop it further.

STEP 3

During the Step 3, Feasibility Analysis, a feasibility plan is developed to examine in more depth the innovation's commercial potential. Feasibility plans evaluate major areas including:

- the management team;
- a technical analysis of the innovation and proprietary aspects;
- regulatory and reimbursement issues,
- market growth;
- competitive analysis;
- pricing;
- sales estimates;
- financial projections; and
- paths to commercialization.

Experts recommend having an independent business professional with extensive industry experience review the feasibility plan to provide objective feedback about its commercial potential. For example, if the idea is opening a new women's clinic, talk to physicians who specialize in women's health, someone who is familiar with starting a new women's clinic, or others who have experience in this industry. Successful innovators surround themselves with seasoned business professionals and savvy entrepreneurs who have industry knowledge and strong professional networks.

The innovator may be required to present the feasibility plan to decision makers. The goal of this presentation is to provide decision makers with enough information to make an informed go/no-go decision. Decision makers decide whether the innovation should be moved forward, declined, shelved, or continued to be improved overcoming weaknesses.

STEP 4

If the innovation is approved, it proceeds to either Step 4A: Commercialization Strategy or Step 4B: Business Plan which depends on the innovation's commercialization path. If the new idea does not involve starting a new business, the innovation moves to Step 4A: Commercialization Strategy. During this step, the innovation is sold, a licensing opportunity is identified, or a strategic alliance or joint venture is formed. Likewise, the new opportunity may simply be implemented within the organization.

If the innovation proves feasible and the innovator wants to start a new venture, it moves to Step 4B: Business Plan. This is where a detailed business plan is written to launch the new venture. The good news is that once a feasibility plan has been completed, about 75 percent of the business plan has already been written. Major sections of a business plan typically include:

- management and organization;
- product/service plan including proprietary aspects;
- reimbursement issues;
- industry trends;
- market penetration;
- pricing;

- financial projections;
- operating and control systems;
- growth projections; and
- risk analysis.

Similar to the feasibility plan, experts recommend having the business plan reviewed independently by a business professional to obtain objective feedback. Once the business plan is presented to decision makers, they will decide whether to put the business plan in action, put it on hold, or decline it.

The time required to move through the Life Science Innovation Roadmap™ can vary from six months to several years depending on the opportunity. The goal is to move an idea through as quickly as possible without sacrificing the necessary due diligence, market research, and validating key assumptions.

Learning how to take new ideas through the steps in the Life Science Innovation Roadmap™ is a career building step. Unfortunately, many innovators are naïve about the commercialization process, how to solve patient problems, develop market-focused products/ services, and leverage the value of new opportunities. Overall, they underestimate the complexities and intricacies of the business development process which can be easily learned.

As Albert Einstein so eloquently stated, "Vision without execution is a hallucination." Creating new ideas is easy and only about 10% of the future work. Taking them all the way to the marketplace is, for most, a hard and bumpy road to navigate with many side detours.

A good resource for writing healthcare feasibility and business plans is "The Medical Practice's Ultimate Feasibility & Pre-Planning" book as well as "The Medical Practice's Ultimate Business Planning" book both with templates and published Medical Group Management in 2008.

IN SUMMARY

Every bioentrepreneur and organization is running the same race — be first to patent, first to market, first to commercialize. Best practices recommend the following when pursuing a commercial opportunity.

- Evaluate an innovation's commercial potential early and often. As new information is gathered and the innovation develops, its potential for success can change quickly. Less time, money, and fewer resources are wasted when weak innovations are abandoned early in the business planning process.
- Quickly assess opportunities using a quick innovation assessment tool or an online assessment tool to determine if the innovation has true commercial potential.
- After quickly assessing the innovation's potential, develop a concise feasibility plan to uncover fatal flaws that must be addressed to improve the innovation's attractiveness and commercial potential.
- Develop a business plan only to demonstrate the viability of a business built around the innovation. If the optimal commercialization strategy is to license the innovation, sell it, or form a joint venture or strategic alliance, there is usually no need for a business plan.
- Validate all assumptions including those about potential customers, IP protection, competitors, marketing, financial, and the technology. Both primary and secondary data should be used to validate each assumption to help reduce risk and improve the chances of commercialization. This process produces solid evidence about the innovation's potential rather than relying on hypotheses and guesswork.

Chapter 4:

Generating New Ideas

Although the world is filled with new exciting ideas, you do not necessarily need a great idea to find a new business opportunity. If you wait until the "Ah-ha" moment to strike, it may never happen. Having the best new idea is not the most critical element. Instead, discovering a new unmet patient need that matches your healthcare organization's or business's strategic plan and capabilities is far better than discovering a new life-saving idea that will so revolutionize health care that there is no immediate need for it or no one is willing to pay for it.

There are two basic entrepreneurial approaches available for generating new ideas within life science organizations. One is an innovative service extension such as adding an ancillary service for additional revenue (*e.g.*, a medical spa). Be aware that not all extra or new services succeed. Sometimes competitors do it first and/or better, or the new idea was just a fad and quickly becomes less popular.

The other approach is discovering a new bioscience invention such as a new medical device, diagnostic, or a drug. However, these types of innovations usually require:

- obtaining intellectual property protection,
- finding short- and long-term capital,
- making a commitment to commercialization, and
- either starting a new business around the invention, licensing it, or finding a strategic partner.

Commercial ideas like these are most often fostered in hospital and/or academic settings because of monetary incentives for the inventors and access to research and development programs. Regardless, all too often the inventor and/or administrators fail to properly assess or vet the feasibility of a new bioscience idea and/or do not develop a sound business case for it. Therefore, ensuring commercial success requires answering such critical questions as:

1. Does the idea satisfy the business and marketing goals for generating revenue or cutting costs?
2. Do we want to penetrate a certain market?
3. Do we expect to increase our payer mix?
4. Do we want to be the first to use a new gadget?
5. Does it align with the goals and culture of the organization?

Following the Life Science Innovation Roadmap ™ requires developing a feasibility plan to determine if the new idea could be successful, and if so, then writing either a commercialization plan or a business plan to develop and launch it. Often healthcare organizations and bioscience inventors are too focused on the short-term revenue and are uncertain how the new invention fits into the big picture and the organization's long-term outlook. The commercialization process may also be flawed because some healthcare organizations fail to offer incentives to physicians and medical professionals for generating new ideas. In addition, they require innovators to jump through all the hoops involved with planning, developing, and commercializing a new invention. In contrast, many universities provide incentives for inventors and frequently assist in helping to commercializing their inventions.

Every discovery should align with the healthcare organization's strategic objectives. Sometimes the organization lacks strategic objectives so there is little or no direction or motivation to do business differently. For example, a physician employed by a hospital goes to a trade show, sees new technology already being used in Europe, and wants to be the first to use it and build a service around it in the

US. An obstacle now appears. Getting decision makers to buy-in and support this new idea is usually very hard.

The planning process at this stage is vital. Getting clinicians to want a new invention or service line and paying for it is critical. The physician or medical professional with the idea rarely has control over the available resources so he/she needs to recruit others' support. During the idea assessment and feasibility analysis stepson the Life Science Innovation Roadmap™, innovators begin to realize what will or will not work. This is the time to identify faulty concepts and determine whether the invention can be maturated or should be dropped. Without developing a feasibility plan, a weak idea may get funded and implemented thereby draining the healthcare organization's precious resources. No one wants to kill it once it is operational, has been staffed up, and resources have been allocated. This is also the time to spot the ideas with a realistic chance of success and worthy of going on to the commercialization stage.

TEN HEALTH CARE IDEA GENERATORS

There are ten good approaches to discover new health care business opportunities which are outlined below. Each has its own characteristics, problems, and probability of success. For example, if you decide to launch a new venture, you undertake a series of new start-up activities. If you decide to buy an existing health care company, another set of activities is pursued. If you plan to commercialize your new invention with your employer, different steps are required.

Review these ten key idea generators to help you identify and capitalize on new business opportunities to take to market. They will help you focus on potential new opportunities that match your interests and strategic initiatives.

1. CUSTOMER-IN-HAND

You do not have a solid business opportunity until you have a patient, provider, or a supplier as a customer who wants what you have to offer. Potential customers who need your product/service are usually the focal point for any new health care opportunity. For example, a patient might contact a clinic and request a particular service that the clinic does not currently offer. Upon further research, a clinician determines that other patients have this same need. Additionally, a group practice may have a need for a complimentary service to one of their treatment methods. This group practice may be willing to be the first customers of the service. This is an example of the customer in hand strategy.

The customer-in-hand strategy reduces many risks inherent in pursuing new opportunities. Whether a new idea will interest enough potential customers to be profitable is often uncertain. Identifying potential customers who are eager for the new idea is the ideal situation.

2. OBSERVE PATIENT'S NEEDS

Your current healthcare employer or life science business provides fertile ground for discovering new business opportunities. Focus on unmet patient needs and needs of other health care professionals who are frustrated because their needs are not being met. Direct involvement with patients on a daily basis can uncover some burning needs and can provide a unique perspective as to how to design a new product/service based on those unmet needs. If you have less interaction with patients and other customers, listen to unmet needs from sales representatives and other frontline staff.

Observe what is important to your patients or customers today. Determine what are their problems and priorities. Figure out if you can do something better, quicker, or cheaper. Innovators are problem-oriented and good listeners. They uncover ideas that will improve their patients and customers' lives. In addition, be sensitive to the needs and changes occurring in the health care industry.

3. TECHNICAL CAPABILITIES

You can use your technical skills as the basis for launching a new business idea. Many people who were educated in the life sciences have discovered breakthrough molecules and then marketed their inventions. A good example is a physician who was interested in bioengineering invented an easy-to-use, minimally invasive handheld device for medical and dental providers to optically detect oral cancers. Identifying potential synergies within your skill sets is a great way to find new business opportunities.

4. INVENTIONS

Sometimes new inventions provide fertile ground for new business opportunities. On the other hand, because of the high cost of research and development, tough regulatory requirements, and the extreme difficulty in marketing them, inventions may make poor opportunities. The track record for taking new inventions to the marketplace is not great since usually the invention is very early stage, the market has not been proven, it cannot be massed produced, and/or the market-penetration strategy is ineffective. For these reasons and the expense associated with protecting and marketing inventions, few are successful. For every iPod, there are thousands of inventions that do not live up to their expectations and lose money.

On average, only two to four percent of all inventions ever earn money. Health care laboratories are full of all sorts of technical gadgets and biomedical devices that do not work well outside the lab. It is a long trail from an invention's prototype to a workable, commercially feasible product. A lot of time, money, and effort are spent between these two points. In addition, many physician inventors have exceptional technical expertise and are excellent designers and idea people, but often lack a working knowledge of customer needs and competitive products and do not have experience in marketing new inventions.

The money and time required to bring inventions to market and to a point of profitability may require many years. The original idea for television can be traced to the years after the Civil War. Considerable money was spent on its development in the 1920s, culminating in its public display at the 1939 World's Fair in New York. Still, profits did not come until the 1950s.

New inventions can be quickly copied and reengineered if they prove successful. Lestoil, the first all-purpose, liquid cleaner, uncovered a market need when it was first introduced by a small firm. Then Procter & Gamble introduced Mr. Clean™ and captured most of Lestoil's market share. Even so-called patent protection proves illusory, as people are able to design around the patent, ignore it, or successfully challenge it in court.

Becoming educated about the commercialization process is the best way to turn your idea into a money making invention. Only a small percentage of inventions eventually make money. But when and if they do, the returns can be great.

5. NON-TRADITIONAL MARKET ADAPTATIONS

Questioning the status-quo within a market is an excellent way to uncover potential innovations. Look at suppliers, competitors, and complimentary service providers and ask why they do what they do, and why your organization has to rely on them. Are there items in your industry that could be brought in-house?

Oftentimes patients find it difficult and unreasonable to go to multiple locations to get tests or to receive the proper treatments. Research services historically offered by competitors and evaluate if there might be new opportunities to explore. Such diagnostic services might include MRI, CT scan, cardiac cauterization, nuclear medicine, freestanding diagnostic imaging centers, investment in ASCs, in-office and independent clinical labs, emerging treatment methods/procedures/products/services consistent with provider specialties, and clinically indicated; *e.g.,* bariatric surgery, cosmetic services, vein clinics. Are there services traditionally offered by competitors that your organization could offer?

While reviewing market adaptations, research the external factors that cannot be controlled such as third-party payer reimbursement, demographics, economic conditions, government regulations, and competition. There are corresponding internal factors that also cannot be controlled such as the current quality of care and patient services, operational conditions and efficiency, access and willingness to devote capital to new ideas, and current management expertise.

Potential synergies are often very difficult to identify. Many times synergies which appear excellent on paper do not work in actuality, and there may be a valid reason for why things are done the way they are. Overlooking this factor could be fatal. For example, NPS Pharmaceuticals realized they should outsource the manufacturing of their drugs to reduce costs. Originally they believed they were saving money by internally manufacturing. There were other companies with better manufacturing capabilities that could manufacture the drug much more economically. The original assumptions about manufacturing costs turned out to be inaccurate.[1]

6. ATTEND A TRADE SHOW

Be on the lookout for new business ideas at trade shows and professional conferences especially in your geographic area. They are a great way to collect updated information about new trends, products and services. For example, most research shows that about 50 percent of people attending trade shows do so to see new revolutionary products and services. You may be able to identify new opportunities that fit your interest and skills.

However, many innovations can come from outside your field so plan to attend trade shows, conferences, and professional meetings tangential to your field. As an example, electronic medical records and telemedicine originated from aerospace programs and discoveries.

Attend local meetings and events with venture capitalists, angel investors, or business development professionals. In addition, read several trade magazines and observe what other health care providers are doing as well as track interesting innovations in other industries. Blogs and online news forums are a great way to keep up-to-date on hot industry topics.

7. BUYING AN EXISTING BUSINESS

There are advantages when buying an existing business which include a proven location, fewer start-up pains, and better financing options which are summarized below.

A. Proven location

Sometimes the only way to obtain a good location is to buy an existing practice or business that has the desired location. It is not so much buying a business as buying a place in which to do business. This works best for medical practices where there are strategic advantages to location, but there also may be some benefit to biotech or

1 Geller, L. and Rotz, G. (2010) "Getting big by going small" *Strategy + Business*

pharmaceutical businesses. Look at tax incentives and proximity to suppliers and customers.

B. Fewer start-up pains

A great deal of time, effort, and money are involved in the start-up phase of new ventures. It takes time to get a new venture rolling. Many of these problems can be avoided by buying an existing business. It is true that other problems can be inherited, but at least there is an existing business with revenue flowing from an established clientele. There is also an experience-base upon which to expand. Starting a venture from scratch may take three to five years to get to the point where an existing business is today. So buying an existing business can reduce start-up time by three to five years.

C. Better financing options

When buying a venture, it is easier to find low cost financing; that is from the seller. Sellers, in their eagerness to make a deal, often find buyers with whom they have confidence and may offer highly advantageous terms of sale. The deal allows the venture to buy itself — the buyer pays the seller from the practice's operating profit which is a great advantage.

Acquiring any new business requires much due diligence to ensure that a purchase is synergistic and will bring value. For example, it is a quicker way to expand a medical practice versus using internal, organic growth. However, there are also challenges with blending cultures, consolidating services, possible staff lay-offs, and so on. Many mergers and acquisitions just never work and cause more headaches than bring in new revenues.

8. CAPITALIZE ON NEW TECHNOLOGIES

Many patients turn to the internet to research their physicians, adding to the importance of well-organized website with high quality information. Many healthcare providers use the internet to schedule appointments, fulfill requests for prescription refills, and facilitate easy communication. Some healthcare provides have experimented with social networking sites such as Facebook to make it easier for patients to refer their friends to a provider. Some have produced fan pages to create some buzz in their local communities about a provider and to provide information about the business to potential clientele. This could be one way to receive feedback about the provider by asking "fans" what they think of the organization as well as to look for innovative ideas from patients and other consumers.

There have been significant advances to digital healthcare, and this utilizes things like mobile phone technology, and web-chat. A good example of this type of innovation is from developing countries where necessity is breeding innovation. In Mexico, MedicalHome offers health advice to subscribers for an additional fee on their monthly phone bills.

This technology could be adapted on a smaller scale to provide an additional service to clients, and possibly determine whether or not they should come in for a visit depending on the illness. This type of lower cost care could broaden the number of clients served by practitioners, and hopefully increase revenues. Keeping aware of new technologies and thinking of potential uses for them in your organization is a great way to generate innovation.[2]

9. NEW WAYS OF DOING BUSINESS

New healthcare business models are being created such as concierge medicine and new medical payment systems. Concierge or boutique medicine where the patient pays an annual fee for enhanced primary care as well as dental and other care is becoming more popular. There are also new medical payment systems being developed, such as episode payment, where a patient pays a single price for all the healthcare services needed for an entire episode of care. For example, a patient pays one price for all inpatient and outpatient care needed after a heart attack. Another new business model is comprehensive care payment where a patient pays a single price for all services needed by a specific group of providers for a set period of time. Many new opportunities can be generated from finding a new and better way of doing business. As professor and author of *Competing for the Future*, Gary Hamel states, "Competition today is not between different products — it's between different business models."

2 Ehrbeck et al, (May 2010) "The emerging market in health care innovation" *McKinsey Quarterly*

10. HOT CURRENT TRENDS

Examining some external trends to determine if there is a potential for capitalizing on emerging trends such as continuing globalization, new technologies, new uses for and advancements in information technology and biotechnology can help generate new ideas. Some of these hot new trends are briefly discussed below.

A. Globalization

Globalization is a continuing trend to discover that new opportunities abroad. For example, more than half of the market for technology products is outside the United States. These products include medical devices, generic drugs, healthcare IT applications and infrastructure, and mobile health. Global expansion is occurring because of falling trade barriers, improved communications, expanding markets in developing countries, increasing global wealth, and the Internet.

The globalization trend can benefit the healthcare and the life science industries. Medical tourism is an example of how globalization is affecting the life science industry. Many residents now leave the US for medical care, and while there is the obvious loss of a patient in the US, there are opportunities to use this globalization factor to your benefit. For example, domestic organizations partner with foreign organizations to offer the pre- and post-treatment care for procedures performed abroad.

Especially in the life sciences, considering the fact that operating internationally requires twice the regulatory work and something that is legal in the U.S. may not be legal elsewhere is important. For example, the Cleveland Clinic opens, staffs, and manages clinics in other countries. The Cleveland Clinic determined the only cost effective way to reach out to their international patients was by contracting to manage clinics and hospitals overseas. They currently have operations in Abu Dhabi, Vienna, and Toronto.[3]

---

3 Buescher, B. and Mango, P. (2008) "Innovation in healthcare: An interview with the CEO of the Cleveland Clinic" *McKinsey Quarterly*

B. NEW USES FOR THE INTERNET

The internet has opened up a lot of doors for the life science industry and also created a few problems. Most physicians are aware of sites like WebMD.com which allows patients to research symptoms before coming to the office, but it also causes patients to come to physicians thinking they have the direst of illnesses.

Cloud computing is an area of growth that could be leveraged by life science organizations. It allows data storage on community servers, and access to the information is available remotely from any device with Internet access. Google Docs™, for example, uses the cloud so individuals can store and share documents. With proper security protocols an organization could store patient records or other important data and only pay for the space that is used.

Leveraging the data that can be captured from the internet could generate an exciting medical idea. Google Analytics™, a data analysis tool offered, charted the outbreaks of the flu and other diseases. It found that the symptoms of the outbreaks were Googled about two weeks before the outbreak was recognized.

The internet can also change how patients interact with healthcare providers. One such way is simply talking to patients via email which saves them a visit to the office for the more simple issues. Video conferencing can allow providers to see the patient or a wound before a patient arrives at the facility allowing the provider time to prepare for the specific issues.

C. NEW TECHNOLOGIES

The most powerful force shaping the way people work, live, and communicate is the evolution of technology. This is especially true of life science organizations that rely on advances in technology to improve patient care and data management.

Even if an organization is already keeping up on the latest healthcare trends in technology there is always the potential to take existing technologies intended for other industries and applying them in the healthcare industry. Adapting technologies from one industry from another can be challenging, and there should be an emphasis on the feasibility planning to ensure that the technology can be a profitable innovation for the organization.

D. Information technology advances

Information technology (IT) is a hot opportunity area for life science organizations. It improves competitive advantage by providing solutions that can be applied to value, cost, lead time, and helping achieve uniqueness. IT solutions and services can provide organizations a competitive advantage, increase intellectual property assets, and give value to shareholders. Many organizations are investing in tablet technology to have portable access to information. A key aspect to focus on when integrating IT into a life science organization is that being on the cutting edge of IT may not always be necessary.

That being said, new IT investments can help a life science organization be competitive, save more lives, and even reduce costs. Another way to use IT in the life science industry is to reduce the inaccuracies that can cause catastrophic mistakes. Technology does sometimes fail, but most automated systems typically produce much fewer errors than humans.

E. Biotechnology

Many biotechnology opportunities exist in agriculture, landscaping, and food-related industries. In addition, there are many new drugs and therapies focused on life extension and improving the quality of life. Recombinant DNA and genetic engineering, for example, have great potential in the areas of preventive medicine. It will be possible to head off chronic diseases by identifying genetic propensities and take early action by making lifestyle changes. Life expectancy will continue to increase, creating many new markets for our aging population. The following are some of the opportunities in biotechnology:

- Blood tests to screen for genetic diseases,
- Genetic engineering,
- Alternative medicines,
- Bionic parts and artificial organs,
- Bio-electronic products,
- Bioengineered crops,
- Animal drugs,

- Inoculated eggs,
- Implantable animal microchips, and
- Grass that doesn't need cutting.

F. Prepackaged services and standardized processes

For years, many physicians have offered ancillary services and products. For example, some ophthalmology practices offer eye glasses. More recently, many dentists have begun offering cosmetic dentistry as natural service add-on. Prepackaged services add value to any healthcare organization. In addition, patients often are willing to pay out of pocket for extra services.

Some organizations have implemented a standardized system of care for common illnesses or issues its patients face. Many of a life science organization's costs stem from the fact that every care plan must be individualized and constructed as needed. Aravind Eye Care Systems has a streamlined, standardized process from operation to recovery to save costs and keep the patient informed of what is ahead.[4] Life science organizations can benefit from standardizing processes that do not need to be specialized for each patient.

G. Medical spas

In addition to offering the traditional spa services, *i.e.* facials and massages, many spas are now adding complementary or alternative health services such as microdermabrasion, chemical peels, and tattoo and hair removal in a spa-like setting where the staff is supervised by a licensed health care professional. As another customer offering, some medical spas have developed its own line of skincare products that generate additional residual revenue.

H. Wellness programs and clinics

Many people are willing to invest their time and money in lifestyle programs including health and fitness assessments, personal nutrition counseling, strength training, and exercise instruction. This includes weight-loss clinics, corporate wellness programs, and longevity centers to name a few. For bioentrepreneurs, programs with a

4 Ehrbeck et al, (May 2010) "The emerging market in health care innovation" *McKinsey Quarterly*

lifestyle focus can open a window of opportunity.

I. Retail and quick-stop clinics

Today's more sophisticated consumers realize that many types of ailments can be treated satisfactorily by nurse practitioners or physician assistants. There are several reasons why retail clinics appeal to many different consumers: accessibility, availability, accountability, accommodation, convenience, and affordability.

There are retail cosmetic clinics sprouting up all over the country that offer varies types of cut-rate, minimally invasive cosmetic surgery in local shopping centers and malls in under two hours.

STIMULATING NEW HEALTH CARE IDEAS

If your organization is interested in stimulating new health care opportunities from staff scheduling a brainstorming session at annual strategic planning retreats and/or appointing an innovation committee.

RETREAT ACTIVITY

A good way to get everyone involved is to have an open discussion of potential revenue-enhancing initiatives during a strategic planning retreat conducted by a trained facilitator. A facilitator could use the outline below to generate a list of new health care opportunities.

- list all options that come to mind that have occurred to the physician and other health care professionals as they follow trends via trade shows and industry publications;
- list services traditionally furnished by hospitals such as: diagnostic and laboratory services, and emerging treatments consistent with provider specialties and clinically indicated; *e.g.*, bariatric surgery, cosmetic services;
- draw initial consultations about the feasibility of these options in light of external factors that cannot be controlled (*e.g.*, trends related to payer reimbursement, population demographics, area economics, likely

competition) and also in light of internal factors (*e.g.*, operational conditions, willingness to devote capital to a new venture, management expertise)

- seek consensus on several desirable innovations; and
- prioritize the list resulting in no more than three options to investigate via a feasibility plan, including estimating revenues/costs. Present conclusions/ recommendations to the group.

APPOINT AN INNOVATION COMMITTEE

New business ideas do not just pop up because a healthcare organization decides it wants to generate additional revenue. Dedicating resources to motivate the entrepreneurial spirit of physicians, non-physician providers and administrative staffs critical as well as listening to needs/wants noted by patients and payers. One way to get the ideas flowing is to form an Innovation or Business Development Committee. It would be responsible for collecting new business ideas, prioritizing them, and performing due diligence related to the feasibility of the most before proceeding into formal business planning.

Such a committee can set the direction for revenue enhancement and diversification. They could be responsible for doing a financial, legal and operational analysis which would include estimating revenue/costs, cost/service, impact on operations, opportunity costs, referral patterns, medical community cohesion, hospital relationships, patient perceptions of group, community in general.

REQUIRED TRAVEL AND IDEA GENERATION COMPANY-WIDE

Sometimes getting staff out of the normal routine and exposing them to different experiences can be a great way to spur innovation. The Cleveland Clinic requires every doctor to take at least one trip to learn something every year.[5] Allowing them to choose the location and topic allows the company to diversify the experiences and possible innovations. This could also be scaled down for organizations with less cash, and the requirement could simply be to attend a trade

5 Buescher, B. and Mango, P. (2008) "Innovation in healthcare: An interview with the CEO of the Cleveland Clinic" *McKinsey Quarterly*

show outside of the medical field, or as simple as spend a certain amount of hours doing research for the purposes of innovation. This exposes the organization to innovative thought and also is a tangible way to begin creating a culture of innovation.

SUMMARY

- Generating new ideas within a life science organization or independently is not easily done.
- By using the idea generating tips in this chapter, bioentrepreneurs can discover hundreds of new ideas.
- Become a trend watcher and focus on medical problems that need to be solved.
- Generating new ideas is only the first step in the Life Science Innovation Roadmap™.
- Continually assessing an idea's commercial potential and getting it commercialized is the biggest challenge.
- Not all ideas warrant spending lots of time, effort, and resources so evaluate and vet a new idea's commercial potential often.

Chapter 5:

Opportunity Assessment

The old adage that, "they can't all be gems" holds especially true for life science innovations. Coming up with great ideas is only a part of the process. Unfortunately, the road to commercialization is often fraught with things that can derail even the greatest of ideas. When there is a culture of innovation within an organization there will often be many ideas and new technologies that could add value and be commercialized. For instance, clinicians frequently solve patients' problems by inventing a new device, treatment, and so on. Consequently innovation bounds in the life science industry. But capturing and commercializing or moving an innovative concept from the idea stage to the marketplace is the hardest and most challenging part.

Many new life science innovations are intriguing and novel, but may not be able to be reduced to practice, do not address a significant market need, cannot be mass produced, implemented cost effectively or, cannot be implemented at all. Oftentimes they are not initially evaluated for commercial potential. Technology developers and organizational decision makers tend to focus on an invention's basic technical attributes which are the foundation for potential successful applications. However, the value of these technical attributes based on the monetary potential of the application, the uniqueness of this technical solution, and whether the technology can be protected is neglected. Thus, only a cursory examination of its commercial strengths and weaknesses is assessed. Also, inventors and decision makers most often do not have a clear understanding of the processes and requirements that will ensure success.

One approach to overcome these problems is by evaluating new technologies and assessing commercial potential early on. Early evaluation of commercial potential enables new technologies to be developed in ways to enhance its chances of market success. It helps decision makers choose whether to continue developing a technology and expend the time and effort necessary to obtain intellectual property protection for the technology. It also helps identify businesses or industries that may be interested in, or best suited to, further developing or adopting the new the technology.

THE 14-STEP HEALTHCARE TECHNOLOGY OPPORTUNITY EVALUATION

The 14-Step Healthcare Technology Opportunity Evaluation tool is a simple, quick method to scrutinize an invention with a market and customer focus. It gives a bare-bones glimpse into the invention's commercial potential rather than relying on intuition, a technical assessment, and comments from colleagues and friends. Inventors and organizational decision makers can use it to better differentiate a high potential technology from a weaker technology. It also helps to jumpstart the business planning process by critically examining areas of weakness as well as potential for success. The 14-Step Healthcare Technology Opportunity Evaluation allows you to quickly determine whether a life science invention is worth pursuing any further.

Completing the following 14-Step Healthcare Technology Opportunity Evaluation helps reduce the risk of pursuing a technology with no chance of commercial success and can help bioentrepreneurs decide whether to continue to develop the new technology, shelf it, or drop it. The ratings provide insights into how to improve the technology and reduce inherent risks associated with all new healthcare opportunities.

ONLINE ASSESSMENT TOOLS

If after completing the 14-Step Healthcare Technology Opportunity Assessment you decide to continue developing the new technology a more in-depth evaluation of the technology opportunity should occur. Using an online diagnostic assessment tool can help further explore whether the technology has high commercial potential.

For example, VentureQuest has an online Healthcare Technology Assessment Tool to assess commercial potential and identify specific areas of strengths and potential fatal flaws. The company customizes the diagnostic tool for its clients to use in their commercial assessment process. For additional information refer to www.venturequestltd.com.

PatentCafe.com offers an Invention Evaluation to assess an invention's potential profitability and marketability. This tool focuses on patentable inventions and is available online at www.patentcafe.com.

SUMMARY

- Bioentrepreneurs typically become very excited about their discoveries making it difficult to recognize fatal flaws in the technology.
- Perform a preliminary assessment of a technology's potential for commercialization which prevents inventors and organizations from wasting time and/or ignoring potential pitfalls in a technology.
- Use the 14-Step Healthcare Opportunity Assessment early and often to gauge commercial potential of new technologies.
- Keep in mind is that just because it does not make sense to further develop a new discovery does not make it a worthless invention.
- Separating personal feelings about a technology's commercial potential by doing an objective assessment early on is extremely important.

14-Step Healthcare Technology Opportunity Evaluation

Evaluate a technology by using the rating scale below. Indicate how this technology compares with the fourteen characteristics of the most successful commercialized life science technologies. Do not select a rating based on what could be achieved with further development. Rather, select the number that represents where the technology is today. At this early stage in the assessment process, all the information needed to accurately answer each characteristic may not be available. Thus, select a rating based on the information currently available. After rating the technology, add up the scores to decide whether to pursue the technology further.

To get a more objective view of this technology's commercial potential, have other team members, staff, or advisors to rate the new technology. Compare these ratings and discuss differences in rater's scores and perceptions. Use the scores to develop an action plan of next steps and what needs to be done to strengthen the technology's commercial potential.

____ 1. Stage of Development
Hint:What is the technology's current stage of development?
1 = early stage, researching the opportunity; 5 = ready to launch

____ 2. Technology is workable and feasible
Hint:Does the technology work the way it is supposed to? Does the technology produce the intended outcome?
1 = unproven concept; 5 = proven concept with workable prototype

____ 3. Value proposition
Hint:How much value does the technology offer customers?
1 = unclear or not obvious value offered to customers; 5 = strong, obvious value offered to customers

____ 4. Defined market
Hint:Does the technology have a clearly recognized and defined market?
1 = no recognized or new market; 5 = well established, defined market

____ 5. Customers
Hint: Are there customers ready to purchase the technology?
1 = no existing customers; 5 = customers in place

____ 6. Competition
Hint: How many competitive technologies are currently available?
1 = many competitors; 5 = few or no competitors

____ 7. Revenue generation
Hint: How long until revenue can initially be generated?
1 = more than 5 years to generate any revenue; 5 = less than 1 year

____ 8. Required funding
Hint: How much money will it take to get your idea to the marketplace?
1 = more than $100,000,000; 5 = less than $500,000

____ 9. Intellectual property protection
Hint: Which IP rights have been obtained?
1 = no intellectual property rights; 5 = have obtained patent, copyright, trademark, and/or trade secret protection

____10. The team's entrepreneurial experience
Hint: How much entrepreneurial experience does the team have?
1 = little financial, marketing, and new product development experience; 5 = considerable financial, marketing, and new product development experience

____11. High repeat sales potential
Hint: How often will customers purchase the technology?
1 = one time purchase; 5 = purchase every use

____12. Market ready
Hint: How long until the technology is ready to begin marketing?
1 = two years away from being market ready; 5 = opportunity is market ready

____13. Reimbursement
Hint: What it the likelihood that a third party payer will reimburse at least some of the cost of using this technology?
1 = no likelihood of reimbursement; 5 = 100% likelihood of third party reimbursement

____14. FDA approval
Hint: Will FDA approval be required?
1 = will require FDA approval; 5 = FDA approved or FDA approval not required

____**Total Score**

Evaluation:

63-70 = Excellent opportunity
55-62 = Solid opportunity
47-54 = Opportunity needs some work
39-46 = Opportunity has major flaws
Below 39 = Very early stage: address major flaws or drop opportunity

Go/No-Go Decision

Based on the score, make a decision about whether to continue developing this new technology, shelf it or drop it.

Chapter 6:

Marketing Analysis

Many bio and other entrepreneurs have the misconception that marketing focuses only on sales, advertising and other forms of promotion. Estimating the size of the potential market is the foundation for developing most, if not all, of commercialization strategies.

Marketing is the strategic driver used to position the opportunity in the marketplace and influence the customer or purchaser demand to achieve sustainable competitive advantage. As opportunities move through the Life Science Innovation Roadmap™, life science innovators must constantly analyze current and potential new emerging markets. This includes defining and analyzing the market, profiling potential customers, and developing advantages over competing offerings. Developing marketing plans and strategies for early-stage life science technologies involves analyzing and researching the three primary spheres of marketing as shown in Figure 3.

Market research involves determining if anyone will purchase your opportunity and whether it will fulfill customers' needs or solves their problems. Marketing analysis involves analyzing the market research to segment the market, select target markets, and develop a deep understanding of what makes the customer tick. Lastly, market penetration focuses on developing strategies to attract new customers and retain existing ones to build market share. In this book, this focus is on marketing analysis.

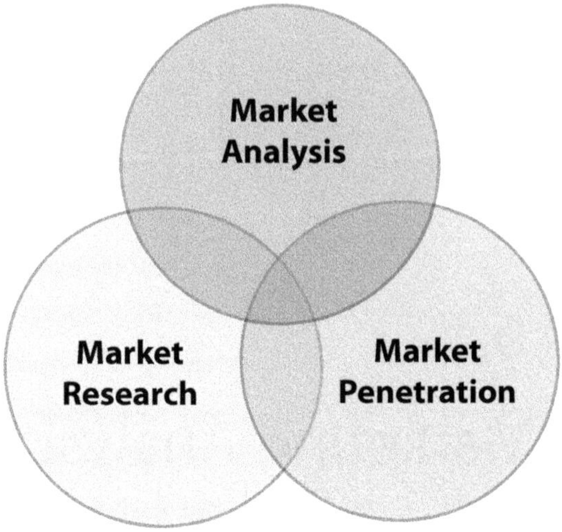

Figure 3: Three spheres of marketing

MARKET RESEARCH

Market research is not dissimilar from scientific research. Rather than developing a new drug or inventing a new device, market research explores whether there is a market for your opportunity. Market research answers the following questions:

- Is there a market for the opportunity?
- How big is the market in dollars?
- Is the market new? stable? growing? or declining?
- Will potential customers and purchasers see value in the opportunity?
- Does the opportunity have a compelling value proposition?
- Does the opportunity meet customer needs?
- Is there a difference between this opportunity and what competitors offer?
- How can the opportunity be differentiated from competing alternatives?

When researching the market, utilize both primary and secondary data. Secondary market data is easily found on the Internet and in magazines, trade journals, books, industry reports, trade association research, and so on. All secondary market data should be validated through primary data.

Collecting primary market research is not quite as easy. It is like clinical trials on humans; it is time consuming and could be expensive to acquire but is absolutely critical. Primary market data can be obtained directly from potential customers, end users, subject matter experts, competitors, or industry influencers. Primary market data gathered from interviews, surveys, and/or observation. Potential investors and/or decision makers will demand primary and secondary data as evidence the opportunity actually has commercial potential.

MARKET ANALYSIS

Markets can be defined and described in many different and equally important ways. They are described in aggregate terms such as "a billion dollar market". Others might describe a market in terms of demographics or psychographics such as 20 percent of the African American population is a market segment. However, these descriptions do not reveal much about an opportunity's potential commercial success. Market analysis helps bioentrepreneurs define and profile the most promising markets and customers with unmet needs who are willing to either use and/or purchase the opportunity. Marketing analysis involves:

- conducting high level market analysis,
- identifying customers through market segmentation,
- determining who pays and who makes the buying decision, and
- understanding the opportunity's features, benefits, and differentiation.

HIGH LEVEL MARKET ANALYSIS

A critical step in analyzing the market is to both define the market and develop a thorough understanding of it at a macro (industry-wide) and a micro (local) level. The macro level market analysis starts with identifying trends that will affect the level of customer demand and the growth rate of a given market. Markets refer to an aggregate group of customers who might buy or use a product/service. Groups of individual customers make up markets and can be subdivided or segmented using demographic (who is the customer) or psychographic (why and how the customer buys) data. Industries, on the other hand, are organizations or companies that make similar products/services for the same markets.

Market trends are typically categorized as geographic (where customers live), demographic, sociocultural (the customer's culture), economic, technological, regulatory, and environmental. For example, a geographic trend is the re-urbanization of America as people move back into cities from rural areas to find work. A demographic trend is the world's aging population. A sociocultural trend is the popularization of organic and vegan diets. Lowered salaries and fewer businesses providing healthcare insurance to their employees is an example of an economic trend. The emergence of electronic medical records is a technological trend. A regulatory trend is the continuing debate over recent healthcare reforms and stronger FDA oversight. Global warming is an example of an environmental trend.

For the life sciences and healthcare industries in particular the landscape is rapidly changing. As discussed in Chapter 1, the primary threats and opportunities involve healthcare delivery and payment reform, an aging population stressing the system, out of control costs, an evolving health care information technology infrastructure, and rapidly changing business models for the biopharmaceutical and medical device industries.

IDENTIFYING CUSTOMERS

One pitfall for bioentrepreneurs is not fully understanding or caring about whom potential customers are. Although in the healthcare industry the real customer is usually the patient, most often it is the healthcare professional, healthcare provider, or insurer that makes the buying decision. This is why many bioscience entrepreneurs are uncertain about:

- to whom they will sell,
- what the purchaser is willing to pay,
- what costs the third party payers will approve,
- what unique customer problem(s) their opportunity solves, or
- which alternative solution(s) is currently being used to solve the problem.

In our competitive and rapidly changing healthcare marketplace, marketing biotechnology has become more and more sophisticated and specialized, often driven by physicians, patients and third party payers. This requires determining who makes the buying decision and what influence patients have, if any, about this purchasing decision.

Physicians are the key decision-makers for many medical devices and drugs as well as for services at hospitals and clinics. As a result, physicians' preferences should be the focus of market research activities since they are probably the key determiners of value. The role of the buyer is also critical as well as the patient, the insured person, the insurer, and the provider.

Becoming familiar with the total market that could buy the solution is a good start. Once the larger market is defined, a subset of different market segments which could be targeted should be selected. This process is called market segmentation. By segmenting a market, different target markets which have customers with differing needs, wants, demands, expectations, preferences, and buying decision processes are identified. A target market is simply a sub-group of customers who have both common characteristics and preferred buying decision processes.

MARKET SEGMENTATION

Segmenting the market allows the bioentrepreneur to focus his/her marketing research and resources on those target markets that are most likely to purchase the solution. Bioentrepreneurs may choose to target one segment, two segments, or many segments. By selecting and prioritizing key target markets, the bioentrepreneur can then develop market penetration strategies and design for effective marketing campaigns for each different target market. In addition, a different product offering could be created for and sold to each target market.

Ultimate success depends on whether early adopters, those who are willing to use the technology before anyone else, will purchase the technology once FDA approval is received. It is extremely difficult to determine the rate at which a new technology will replace other solutions to the same problem.

All successful bioentrepreneurs intimately know who their customers are and what they need. Keep in mind that the end-user of a bioscience product most likely is not the same person as the customer.

Market Segmentation can be done in a variety of methods. The following discusses the main characteristics used to segment markets.

- **Demographic Segmentation**
  This most basic method of segmentation is popular because customer choice of many products and services correlates with their geographic location, age, sex, ethnicity, income, and occupation. The use of demographics to profile customer is a technique used by most marketers.

- **Psychographic Segmentation**
  Psychographic segmentation is much more sophisticated than the demographic approach because it attempts to characterize the psychological factors of personality, attitudes, motivation, perception, and the sociological factors of culture, social class, lifestyle, group influence, and personal and family influence. Psychological characteristics can also be based on activities, hobbies, and recreational activities.

- **Behavioral Segmentation**
  Behavior segmentation concentrates on dividing

markets into situational specific variables such as frequency of use, brand loyalty, and product/service benefits. One type of behavior segmentation is call benefit segmentation. This type segments the market on the basis of the benefits customers are seeking from a product or service. This is extremely useful because bioentrepreneurs can decide which benefits they choose to focus on and can then devise a specific strategy around those targeted benefits. Other behavioral characteristics include tendencies, prices paid, and the purchase/decision process.

- **Geographic Segmentation**
  Segmenting the market into geographic areas seems quite simplistic, but it is very important when defining the target market. With increasing ease, products and services are being delivered to all parts of the world each with different characteristics. Defining market segments in terms of continents, countries, states, and localities can be advantageous.

- **Solution-based Segmentation**
  A lesser-known method of segmentation is based on solutions. This incorporates determining the unmet needs, dissatisfactions, and frustrations that consumers have to define a segment. The more narrowly a target market is defined, the more focused the marketing activities can be.

TARGET MARKETS

Selecting the market segment the innovation is best able to satisfy and sustain a competitive advantage is critical. Understanding that a product or service cannot be "everything to everyone" is an important concept for the bioentrepreneurs to understand. Specialized, focused marketing to a specific target market(s) will help build competitive advantage and allows bioentrepreneurs to better understand their customers and predict their customer's future behavior.

Successful bioentrepreneurs use the following guidelines to profile their target customers:

- Define the characteristics of customers including age, occupation, economic status, ethnic background, size of family, and level of education
- Determine the benefits and expectations that the customer seeks including values such as integrity, reliability, professionalism, and lifestyle
- Identify why a customer would purchase and also why they may not purchase
- Uncover trends in buying patterns and shifts between needs and expectations across demographic segments
- Evaluate the threat of substitute products and how competitors build differentiation using variations in their unique service proposition
- Assess the method of delivery or the channel in which the customer prefers to seek your products or services.

DETERMINING WHO PAYS AND WHO MAKES THE BUYING DECISION

The healthcare marketplace is substantially different from other industries since the person receiving the service, the patient, is often not the person who chooses the service. Instead it could be the doctor and/or the person paying for the service or the payer. In addition, payment is either pre-paid in the form of private or government insurance, self-paid, or paid by a safety-net payer. The marketing decision needs to incorporate how each stakeholder, *i.e.* patients, payers, providers, and partners (industry) will view your product/service, adopt it, and pay for it.

OPPORTUNITY'S FEATURES, BENEFITS, AND UNIQUENESS

As you develop your bioscience opportunity, it is important to first take an in-depth look into the opportunity's features, benefits, and its unique aspects. Developing a features and benefits analysis before you start writing feasibility or commercialization plan will help evolve your product/service idea and determine if potential customers might purchase it over that of competitors. After completing this exercise, you can create key selling points that can be used to convince others to support moving your idea forward. Use this exercise to turn features into powerful and compelling benefits which are used in marketing collateral and sales presentations.

Volumes have been written on the importance of features, benefits, and uniqueness because they form the foundation of marketing. Translating features into benefits and uniqueness is an art and difficult at best. It occurs by:

- translating product/service features into need-satisfying customer benefits,
- identifying and matching features, benefits, and uniqueness to your customers' needs,
- building a strong customer value proposition which enables you to market your product/service more successfully.

Marketing success stems from understanding what your potential customers need and want and then communicating that you have the best solution by highlighting your product/service's unique benefits. Product/service features, benefits, and uniqueness are effective tools to use to get potential customers to fully understand and appreciate your solution's true purpose.

Many people think that customer benefits are the product/service features and what they do. This is not the case. The benefits provided to the customer are the true reason they purchase the product/service. Rather than focusing on what a feature does (*i.e.,* sutures a wound), think of what that feature means to the customer (*i.e.* leaves no scar). Benefits are always written in a language that communicates this meaning at an individual, personal and emotional level. The more intimate your benefits are, the more real, vivid, significant, and meaningful they will be.

FEATURES VERSUS BENEFITS

To generate maximum marketing impact, you first need to understand the differences between features and benefits. Very simply:

- Features are characteristics that physically describe how a product/service works. In other words, features are what products and services *have* (3-D visuals, thin needles, and so on).
- Benefits are what the feature *means* to the customer and describes how your product/service helps them solve their burning problem (reduces recovery time, minimal scarring, and so on).

Features are easier to identify because they are tangible and focus on what the product/service does. Benefits, conversely, are more difficult to discover because they are usually intangible and explain why someone would use the product/service. No matter how good your product or service, if you are unable to tell potential customers why it benefits them, you will not advance your cause. Put simply, you must be able translate your features into benefits.

Unfortunately most bioentrepreneurs focus on the product/service's features and then guess about its benefits and unique characteristics for their potential customers. Examples of product/service features can be easily found in the product description on a company's website, product packaging, or in the technical specifications. Studying product descriptions of similar products/services will help you better define your product/service's features.

The product/service's benefits will explain to potential customers why they should adopt your solution, what they will gain from using it, and why they would want to buy it. Think of what a feature means to the customer and then translate this to the customer's personal and emotional level.

For example, if one purchases this new patient service how will it improve their life? Will it make them feel better? Will it make their life easier? Will it save them money? Creating and selling benefits involves an insight into the potential customer, understanding their motivations, and considering the customer's viewpoint. These kinds of benefits are sometimes hard to discover without talking to your customers. The person buying a boat is not just buying a boat; they are buying a lifestyle.

UNIQUENESS

One of the core aspects of developing a successful product/service is defining its uniqueness from competitive or alternative solutions. Uniqueness is the core differentiator behind successfully commercializing a new product/service. Most likely, there are many competing similar products/services which provide comparable benefits. Therefore to distinguish your solution, you need to precisely identify your product/service's inherent unique characteristics.

Discovering your uniqueness comes from comparing your product/service to your competitors by studying their websites and analyzing their products and services. Look for what makes your solution unique from theirs. If you cannot find any uniqueness, you may need to reconsider your product/service so that it is truly unique and different. If your product does not have any unique characteristics when compared to immediate competitors, it may not be successful in the marketplace. Keep in mind that if there is just one competitor for your product, you need unique characteristics to differentiate your product/service. Examples of uniqueness include: supreme durability, quality, functionality, design, etc. It is your job to make your product/service stands out from competitors and really captivates potential early adaptors.

DEVELOPING FEATURES, BENEFITS, AND UNIQUENESS

Remember, customers relate more closely with product benefits and uniqueness than with features. The most successful marketing campaigns focus on the benefits and uniqueness, not the product features. For example, think of a Snickers commercial. Mars, Incorporated intentionally does not describe the candy bar with chocolate, peanuts, nougat, and caramel; rather, they tell you that it satisfies hunger. The best way to distinguish between features, benefits, and uniqueness is to look at some examples.

	Feature	Benefit	Uniqueness
GE Valid E9™ Ultrasound	One-touch 4-D & 2-D screening	Improves accuracy, saves time	Only ultrasound available that is tailored for each patient need
Crest White Strips	Whiter teeth	Increased confidence, improved appearance, look younger	Removes up to 20 years of stains in less time than others
Balloon Sinuplasty™ Device	Catheter-based device to clear blocked sinuses	Preserves normal anatomy & mucosal tissue	Can be used with other medical therapies or techniques
Syndicom ProductEdge™	Web-based network connecting device companies to surgeons	Accelerates product adoption, increases product use, produces better surgical outcomes	Builds on the success of Syndicom's members such as UCLA & The Cleveland Clinic

Benefits and uniqueness are not just hype and statements full of hot air. Instead, they are effective tools you can use to develop and communicate your customer value proposition.

The Features, Benefits, and Uniqueness Exercise on the follow page helps you further develop your product/service's features, benefits, and uniqueness. The easiest way to discover your benefits is to put yourself in your customer's shoes and ask, "What's in it for me?" for each listed feature.

To get started, go to competitors' websites, review their technical specifications, and list their features and benefits. Next, list your product /service features. Remember that for each feature you list on the exercise below there needs to be an accompanying benefit. If you do not have a companion benefit, your customer's will not care about the feature. Understanding why a feature exists will help you identify its benefits. For example, computers come with a non-glare monitor. Customers benefit from this feature because they can easily see the screen regardless of the lighting in the room. Another simple example is a new car feature is 80 miles per gallon. What is the benefit? The benefit is that you will save a fortune buying fuel while helping to conserve valuable fuel resources.

Make sure that you describe your benefits in a way that is appropriate for, and directly related and targeted to, your customers. We often cannot see the forest for the trees. Instead, focus on how the features you list specifically benefit your customer. Remember, different words mean different things to different people and you want to describe your benefits in your customer's language. Only list those benefits that truly reflect and cater to your customer's problems, needs and desires. It is virtually impossible to successfully complete this exercise without talking to potential customers. Ask them about their needs and perceptions. Test your product/service for usability and desirability. If you have existing customers, ask them about your product/service benefits. Once you have listed benefits that appeal to your potential customers, list the uniqueness of your product/service over your competitor. The only time customers want to hear about your product/service's feature is when it provides them with a desired benefit and is a unique solution. The Features, Benefits, and Uniqueness Exercise helps you articulate your value proposition and learn how to better market your product/service. Remember: features tell, but benefits sell!

FEATURES, BENEFITS, & UNIQUENESS EXERCISE

Instructions: First list the features of your product/service: what makes it work? Next, list a benefit for each feature you have listed in the Features column. Then, address the uniqueness of how it differs from competing products/services. Finally, ask yourself, "Did I list all the benefits a specific feature has? Do the benefits reflect and cater to the needs of my potential customers? Is it different from other products and services? Is the language easy to understand?"

Features What does the product/ service have? What does it do?	**Benefits** Why would someone use the technology?	**Uniqueness** How is it different from other products and services?
Example: *Solar power*	*Economic savings, environmentally responsible*	*No comparable product uses solar power*

Hint: Benefits usually come from the features of the product/service. Remember that customer buy benefits, not features. For example, people don't buy cars; they buy speed, luxury, power, safety, etc. If the feature is solar power, the benefit is economic savings and environmental responsibility. Other examples of benefits are efficiency, time savings, increased safety, increased comfort, more accuracy, cheaper, etc. Uniqueness is what makes your product/service different and better than your competitors. Examples of uniqueness are supreme durability, quality, functionality, and so on.

SUMMARY

- Successful bioentrepreneurs constantly research and analyze conditions in the market and industry at both the macro and micro levels.
- Analyze the market at the macro level to identify trends that either represent new opportunities or threats to your life science innovation.
- Collect both primary and secondary market research to help develop your technology so it is more attractive not only to potential customers, but also to decision makers and potential investors as evidence to back any claims.
- Successful entrepreneurs intimately know their customers including their age, ethnicity, income level, location, buying habits, wants, needs, dreams, goals, and so on.
- Translate your technology's features into customer benefits since benefits are crucial to the customer, pinpoint their pain points, and in effect, solve their problems.

CHAPTER 7:

Protecting Intellectual Property

TO BUILD VALUE around your invention or discovery, you must protect your intellectual property (IP) rights which creates barriers to competition and can also provide a competitive advantage. It is critically important that you, like any innovator, understand intellectual property rights you create from your discoveries. You also should learn about all the options available to proactively protect and manage your IP as well as how to leverage it strategically to generate maximum value for your invention.

Becoming familiar with some basic information about various legal forms of IP is also important such as patents, trademarks, copyrights, trade secrets, domain names, and noncompete/nondisclosure agreements. As a simple example of the differences among patents, trademarks and copyrights, consider the IP aspects of Coca-Cola:

- The artwork on the can or bottle is copyrighted.
- The term Coca-Cola along with the shape of the bottle is trademarked.
- The formula for Coca-Cola is a trade secret but could have been patented.
- Nondisclosure agreements are used to protect the formula as a trade secret.
- Coca-Cola has registered various domain names relevant to its products.

- Many Coca-Cola regional bottlers are assigned the right to geographic regions to limit competition.

How can these intellectual property forms be leveraged strategically? When should an innovator consider patenting versus filing for a trademark or creating a trade secret? When should you consult an IP attorney? Does a copyright need to be federally registered? How much money should be budgeted to protect an idea with IP? Becoming familiar with IP and some of the strategies behind using these legal tools will help you fully comprehend the ramifications of your business decisions. They'll also help you strategically leverage your IP and enhance the value of your discovery.

When dealing with IP issues, the innovator is thrown into the depths of the legal system and most likely will require consulting with an IP attorney. Making small mistakes with IP issues can quickly become expensive, time-consuming, and could destroy the value of your invention. In addition, it is important for Bioentrepreneurs to be aware of and manage their IP legal expenses including fees for protecting their IP. For example, it costs money to file for a patent, copyright, or trademark and even more to fully protect your IP rights. In the United States, the fees for preparation and filing a patent can range from $8,000 to $25,000 or more with additional costs to get the patent issued. It is not uncommon to spend twice the initial cost of getting your patent on file to get it issued and pay maintenance fees. This only protects your idea in the United States, with each other country incurring similar costs.

Keep in mind that it is expensive not only to obtain patents, but also to protect patents. Once legal fees, government fees, maintenance fees, foreign equivalent translation fees, and the value of the inventor's time have been totaled, only inventions with the greatest value are worth being patented. That is why many healthcare professionals working for healthcare providers prefer, when appropriate, to commercialize their IP by working internally with their organization's technology transfer office.

Not only is it expensive to file for your patent rights, but it also expensive to defend your patent rights when others infringe upon those rights. This occurs when a protected item is made, used, sold, offered for sale, or imported into the US without the IP owner's authorization. The US Patent & Trademark Act provides that the owner may sue an infringer and seek remedies of injunction and/or damages in certain circumstances. But it is your or your healthcare organization's responsibility to take action and sue the infringer. Pursuing patent infringements is by far the biggest legal expense for a valuable invention.

For example, Staar Surgical's Coopervision patent infringement case cost the company almost $2 million to pursue the patent infringer. Today, juries in patent infringement cases are increasingly willing to recognize the true value of damage caused by infringement. Damages awards have been increasing since about 1982. From 2000 to 2006, the median damages awarded were about $3.8 million. Many of these damages awards are from medical patents such as an eye surgery procedure and MRI equipment.[1] The largest damages award ever was paid in a patent infringement case involving pharmaceuticals. A Texas jury awarded Johnson & Johnson's Centocor, Inc. $1.67 billion because Abbott Laboratory's TNF inhibitor, Humira, willfully infringed on the patent for Remicade.[2]

In short, intellectual property can be very valuable and should be carefully protected. The more you understand about the IP rights you create, the better position you'll be in to leverage their value and enjoy the benefits for your discoveries. As colleague John Nevard eloquently stated, "The role of Intellectual property has become so pervasive in the business world that it is appropriate to say there is hardly any significant industry that can be classified as IP-free. IP forces are restructuring industries and defining new ways to compete.

Following is a summary of the basic types of IP protection available and a description of each.

---

1 McGrath, Shawn and Kathleen M. Kedrowski. *Trends in Patent Damages.* Section of Litigation, American Bar Association. 2006.

2 Centocor, Inc et al v. Abbott Laboratories, 07cv00139, E.D. Texas (Marshall).

Summary of the Various Types of Intellectual Property Protection

Feature/Type	Patent	Trademark	Copyright	Trade Secret
Intangible asset	Machine, manufacture, process, or composition of matter; business method; software	Brand - identity, source, logos, good name or reputation; trade dress	Original and creative works made up of words, sounds and pictures; software programs	Anything that is kept as a secret and provides a business advantage
Fundamental nature	Structure and function	Source-indicating function	Expression	Confidential relationship
Basic metes and bounds of the Exclusionary (Negative) Right	Make, use, sell, or import by self; inducing another	Use source indicator that is identical or confusingly similar	Copy, publish, distribute, display work, create derivative works	Use and unauthorized disclosure
Threshold requirements	Useful, new, non-obvious, and described properly	Distinctive enough to indicate a single source and used as such	Original with a modicum of creativity	Relatively confidential; useful; duty/relationship
Mechanism for creation	Grant of U.S. and foreign patents	Use of the mark in U.S. and some countries; by grant of registration in most of the world	Creation of an original work in almost every country	Body of laws of each different country

Source: *Intellectual Capital in Enterprise Success – Strategy Revisited,* Dr. Lindsay Moore and Lesley, Craig, John Wiley & Sons, 2008

PATENTS

Gaining a working knowledge of patents is of critical importance to make wise decisions regarding the possibility of patenting products, designs, or processes. Patent law is highly complex but could be critically important to protecting your invention.

A U.S. patent is:

- a government grant of a right to exclude others from any beneficial use of the patent in the United States.
- for a limited term of 20 years from the date on which the application for the patent was filed in the United States or, in special cases, from the date an earlier related application was filed, subject to payment of periodic maintenance fees (a maintenance fee is due 3½,

7½, and 11½ years after the original grant).

- in the language of the statute and the grant itself, "the right to exclude others from making, using, offering for sale, or selling" the invention in the United States or "importing" the invention into the United States.
- not the right to make, use, offer for sale, sell or import, but the right to exclude others from doing so. Indeed, it is common that a patented invention infringes the patents of others.
- available to any person who invents or discovers any new and useful process or method, system, machine or apparatus, or composition of matter [that is, chemical compounds or mixtures of ingredients], or any new and useful improvement thereof, may obtain a patent, subject to the conditions and requirements of the law, and should be able to successfully apply for a patent. This includes software functionality, engineered biologicals and business methods.
- in public domain when the patent expires. Anyone is thereafter is able to use the invention claimed in the expired patent. That is, all of the inventor's or assigned owner's rights are gone and anyone can freely copy the idea described in the patent.

Abstract ideas, laws of nature, and physical phenomena are not patentable subject matter. A patent cannot be obtained upon a mere idea or suggestions alone without some application of that idea. The patent is granted upon the new device, process, molecule, etc., as it has been explained in the description with the claims defining the exact exclusionary right.

There are three types of patents granted by the US Patent & Trademark Office (USPTO). They are:

- *Utility Patents* may be granted to anyone who invents or discovers any new and useful process, machine, article of manufacture, or compositions of matter, or any new and useful improvement thereof. There are three specific requirements for a utility patent:

- The invention must be new or "novel". This means that the invention has never previously been published or made known anywhere in the world. This is why it is recommended that you do a prior art search early on as your begin developing your opportunity if filing a patent is desired.
- It must be "useful" which means that the invention must have some practical value including a useful purpose and efficacy.
- The invention must be "non-obvious," meaning not apparent from the knowledge that existed in the public domain at the time the invention was made. Inventors often find most things obvious in their field, but the legal standard is a far lower bar, so don't discount your innovation before consulting with a patent attorney.

- *Design Patents* may be granted to anyone who invents a new, original, and ornamental design for a functional item. A design patent protects the way a functional item looks. There are three specific requirements for a design patent:
    - The patentable design must be new meaning that the exact design has never before appeared in the public domain.
    - The design must be original or the inventor must have created the new design independently without copying or deriving from another source.
    - It must be ornamental meaning that they patentable design is limited to visual characteristics of the article. The functional aspects are protected with utility patents.

- *Plant Patents* may be granted to anyone who invents or discovers and asexually reproduces any distinct and new variety of plant.

FOREIGN PATENTS

With the exception of certain process patents, a U.S. patent is only good within the United States and her territories and possessions. Items made using a process patented in the United States may be a patent infringement if imported into the United States. In general, a company must obtain separate patents in each nation in which it wishes to have patent protection. A foreign patent application must be filed or based upon a US application that was filed before any public disclosure of your opportunity. When filing a foreign patent off a U.S. patent filing, the foreign process must begin within one year of the U.S. patent's earliest priority date.

PROVISIONAL PATENTS

The U.S. Patent and Trademark Office (USPTO) allows inventors to file a provisional patent application. A provisional application acts as a less-complicated first patent filing in the United States and usually costs less than a full patent application to have prepared. Claims and oath or declaration are *not* required for a provisional application. Provisional applications provide the means to establish an early effective filing date in a patent application and permits the term "Patent Pending" to be applied in connection with the invention. Provisional applications cannot be filed for design inventions.

The filing date of a provisional application is the date on which a written description of the invention, drawings if necessary, and the name of the inventor(s) are received by the USPTO. Once the provisional application is received, inventors have up to one year to file a non-provisional or regular utility patent application. The claimed subject matter in the filed non-provisional application is entitled to the benefit of the filing date of the provisional application if the claimed ideas have proper support in the provisional application.

Provisional applications are not examined on their merits. A provisional application will become abandoned by law one year from its filing date and remain confidential unless a non-provisional is filed. The one-year pendency for a provisional application is not counted toward the 20-year term of a patent granted on a subsequently filed non-provisional application that relies on the filing date of the provisional application. Provisional patents are especially useful when a public disclosure or offer for sale would prevent later filing, but there is no time to properly prepare a non-provisional application.

A brochure on Provisional Application for Patents is available at *www.uspto.gov/web/offices/pac/ProvApp.pdf.*

DATE OF INVENTION AND BAR DATES

In the US, a patent had belonged to the first person to make the invention, not the first person to file a patent application as in most other countries. The law is in the process of changing to follow the rest of the world such that the first to file a patent is the inventor unless the innovation was derived from exposure to another's work. *As the first to invent law goes away, filing patents early in the innovation process will become crucial.*

In addition, US patent laws require that a patent application be filed less than one year from a public disclosure which is called the bar date. This creates a one-year grace period from the time that an invention is used publicly or disclosed on a non-confidential basis anywhere in the world, or sold or offered for sale in the United States to the time an idea has to have a patent application filed. Most foreign countries do not have any grace period such that any disclosure bars patent protection immediately. This is why many inventors consider filing a provisional patent on their opportunity first. The U.S. Patent and Trademark Office's Inventors Assistance Center (http://www.uspto.gov/inventors/iac/index.jsp) is a helpful resource to get questions answered at no cost.

WHEN AND HOW TO USE ATTORNEYS

Patent law is an exceedingly complex field requiring the assistance of lawyers or patent professionals who specialize in IP. A technical background and patent bar exam are required to be a patent attorney. Even some large corporations often do not have dedicated patent attorneys on their legal staff but use outside legal counsel who specialize in Intellectual Property. The USPTO has a searchable database of the attorneys and agents authorized to practice before the PTO (http://des.uspto.gov/OEDC), but no opinion is provided on their work quality. Another suggestion is to ask colleagues for recommendations who have used a patent attorney or firm. Being sure that your patent attorney understands your technology and has the right technical background will go a long way toward controlling costs and assuring a quality job.

HELPFUL RESOURCES

The U.S. Patent and Trademark Office's General Information Concerning Patents should be reviewed on the PTO's website at the following address: http://www.uspto.gov/web/offices/pac/doc/general/index.html.

A book on the subject that may be helpful is *Patent it Yourself*, by David Pressman (Berkeley, Calif., Nolo Press, 2008). Frank H. Foster and Robert L, Shook's *Patents, Copyrights, and Trademarks* (New York, NY, John Wiley & Sons, Inc., 2nd Edition, 1993) is another good book covering trademarks, copyrights, and patents.

TRADEMARKS

A trademark is any word, name, symbol, sound, color scheme, or design or combination thereof, which identifies and distinguishes the sources of goods or services of one party from those of others. Really anything that serves as a source indicator to consumers is suitable for trademark protection. A service mark is the same as a trademark except that it identifies and distinguishes the source of a service rather than a product.

Trademark rights may be used to prevent others from using a confusingly similar mark, but not to prevent others from making the same goods or from selling the same goods or services under a clearly different mark. If you are choosing a mark whose use would confuse consumers of similar goods, another choice is advisable. A trademark attorney can search for similar marks that will cause trouble as you build a brand around your trademark.

If properly maintained, a trademark can last indefinitely; or, in other words, as long as the life of the company, product, or service sold under the mark. Federal registration of your trademark, although optional, will enhance the value of a trademark and serve as a better deterrent to those that might trade on your good name. Your trademark registration requires the owner to file necessary documentation with the USPTO at the appropriate times and pay applicable renewal fees.

Trademarks do not have to be registered with the USPTO; however, owning a federal trademark registration provides several advantages:

- Constructive notice to the public of the registrant's claim of ownership of the mark;
- A legal presumption of the registrant's ownership of the mark and the registrant's exclusive right to use the mark nationwide on or in connection with the goods and/or services listed in the registration;
- The ability to bring an action concerning the mark in federal court;
- The use of the US registration as a basis to obtain registration in foreign countries; and
- The ability to file the US registration with the US Customs & Border Protection to help prevent importation of infringing foreign goods through administrative enforcement.

At any time, trademark rights can be claimed by adding the symbols, ™ or ℠, as designation to alert the public to a trademark claim regardless of whether a trademark application has been filed with the USPTO. However, the federal registered trademark symbol, ®, may only be used on or in connection with the goods and/or services listed in the federal trademark registration.

Infringement on a trademark is actionable in court. In addition, many times, courts will provide protection to an unregistered trademark or a common law trademark under unfair competition laws. A common law trademark is a trademark which has rights that are acquired through the use of the mark in commerce and generally extends the geographic area in which the trademark owner conducts business. Injunctions are often issued where trademark infringement can be proved. Damages and lost profits may also be awarded.

HELPFUL RESOURCES

It is much easier to obtain a trademark than a patent and legal fees are typically far less. It is largely a matter of filing the proper papers and paying the fee. Trademark attorneys can help navigate this process and reduce the risk you are investing goodwill in an unobtainable trademark. Information about how to obtain a trademark can be obtained by reading any of the hundreds books on the subject. The trademark section of the USPTO's website is helpful as well (http://www.uspto.gov/trademarks).

COPYRIGHTS

A copyright is a form of protection provided to author of "original works of authorship" that are fixed in a tangible form of expression. Copyright protection is available on both published and unpublished works. Copyrightable works include:

- Literary works;
- Software code;
- Musical works, including accompanying words;
- Dramatic works, including accompanying music;
- Choreographic works;
- Pictorial, graphic, and sculptural works;
- Motion pictures and other audio visual works;
- Sound recordings; and
- Architectural works.

The 1976 Copyright Act generally gives the owner of a copyright the exclusive right to:

- reproduce the copyrighted work;

- prepare derivative works based upon the original copyrighted work;
- distribute copies of the copyrighted work to the public;
- perform the copyrighted work publicly; and
- display the copyright work or portions thereof publicly.

A copyright protects the form of expression rather than the subject matter of the writing or underlying ideas. For example, a description of a medical device could be copyrighted, but this would only prevent others from copying the descriptions; it would not prevent others from writing their own description or prevent others from making or using the device.

No registration or other action is required by the Copyright Office to secure a copyright. However, registering a copyright has these advantages:

- a public record of the copyright claim;
- fulfillment of a mandatory prerequisite to filing an infringement suit in Federal court;
- evidence of the validity of the copyright and of the facts stated in the Certificate of Registration;
- statutory damages and attorneys' fees in infringement suits; and
- option to record a registration with the US Customs and Border Protection for protection against importation of infringing copies.

Registration can be made at any time during the life of the copyright. Copyright protections subsist from the time the work is created in a fixed tangible form. The copyright in the work of authorship immediately becomes the property of the author who created the work. Only the author or those deriving their rights through the author can rightfully claim the copyright. In the case of works made for hire, the employer and not the employee is considered the author. Consultants, temporary worker, contractors and vendors don't automatically assign their copyrights so the agreement retaining their services should state what rights will flow to the company. A copyright is given a term that typically far exceeds any commercial value so it is a safe presumption that there is a copyright on any material you might encounter. It is not unusual to have the term exceed 100 years and varies by country. The expiration of a copyright should be confirmed by an attorney before undertaking an effort that could result in substantial damages.

Incorporating software code found on the Internet, developed for prior employers or subject to open source license requirements can pollute your code base. Be sure any software is developed by employees or others that properly assign their copyright to the business. Part-time developers and consultants should have a contractual requirement assigning all their copyrights to the business. Do not use code found on the Internet without being sure your license to it is suitable for the software's intended use without excessive license fees. Prior code that an employee or consultant has developed is likely owned by someone else and shouldn't be used without securing the proper license for it. Open source software can accelerate development so long as the business has adopted that strategy which could subject it to providing its code base to anyone including the competition. Scrutinize open source licenses carefully.

FAIR USE OF A COPYRIGHTED WORK

Fair use of a copyrighted work allows for reproduction for purposes in which the reproduction may be considered "fair," such as in criticism, commenting, news reporting, teaching, scholarship, and research. Generally, courts have regarded the following as fair use:

- Quotation of excerpts in a review or criticism for purposes of illustration or comment;
- Quotation of short passages in a scholarly or technical work, for illustration or clarification of the author's observations;
- Use in a parody of some of the content of the work parodied;
- Summary of an address or article, with brief quotations,

in a news report;

- Reproduction by a library of a portion of a work to replace part of a damaged copy;
- Reproduction by a teacher/student of a small part of a work to illustrate a lesson;
- Reproduction of a work in legislative or judicial proceedings or reports; and
- Incidental and fortuitous reproduction, in a newsreel or broadcast, of a work located in the scene of an event being reported.

Copyright infringement involves the copying or reproduction of a copyrighted work without the permission of the copyright owner, before expiration of the copyright term, and not within the "fair use" exemption described above. The copyright owner can obtain damages and an injunction against a copyright infringer. In certain cases where the infringed copyright has been Federally registered, statutory damages may be awarded to the copyright owner regardless of whether the infringement is deemed to be willful or not. In addition, unlike patent or trademark infringement, a copyright infringer may also be subject to criminal prosecution with penalties including fines and imprisonment.

TRADE SECRETS

A trade secret is any valuable information one business knows that others do not know. Trade secrets include techniques, designs, materials, processes, business plans, and formulas that are learned from experience, are not known by the public, and can be licensed in the same way as a patent. Naturally, trade secrets should be something that contributes significantly to the firm's profits or its success. Customer lists, supplier lists, and merchandising or promotional systems can contain secret information and should be protected as trade secrets if the company knows how to do something that others do not. Such trade secrets can be very valuable.

Ideas that are productized can often be reverse-engineered once sold in the market such that trade secret protection is an inadequate protection on its own unless access to the trade secret can be controlled. Other IP should be considered for protection to augment the trade secrets in these cases, for example, patents or copyrights. Conversely, some companies choose to use trade secrets to protect their IP because patents have a limited protected lifespan; and the patent applications require detailed instructions to reproduce the invention which is considered public information. However, it is illegal to sell or promote the reproduced invention until the patent has expired. Consider pharmaceuticals as an example. Once a drug's patent expires, many companies begin to produce "generic" versions of the same drug and sell them for a fraction of the name-brand price.

The courts have been relatively generous in dealing with stolen trade secrets. The law recognizes a trade secret as a company's property and protects it as such. However, the courts have a few requirements:

- The protected knowledge must be secret. If the defendant can show that, in fact, the knowledge is known by others, the court will throw out the case.
- Every precaution must be taken to keep the knowledge secret. Those who are exposed to the trade secret should be notified that the information is to be kept confidential. If information is not protected with the utmost care, the court is likely to consider the protected information to be of little or no value. Courts must clearly see the efforts made to protect the trade secret.
- Employers should place specific provisions in employees' agreements that they will neither disclose to others nor use for their own purposes any trade secrets they acquire while in the firm's employ, and for a reasonable period thereafter.

CUSTOMER LISTS AND EMPLOYMENT AGREEMENTS

Customer lists can be a trade secret and protected. Bioentrepreneurs should take great care in protecting their customer lists from suppliers and competitors. Customer lists are often the most valuable asset of a business. It is critical to use employment agreements which clearly address use of both trade secrets and use of customer lists. Departing employees headed to work for a competitive firm often try to take such information with them. If they manage to obtain a list, they can do much damage to former employers that may be hard to prove in a lawsuit. The courts tend to favor the harmed business, particularly if the act is in clear violation of the employment agreement signed at the time of employment.

Employees should be advised that the information is secret and they should sign a confidential information agreement that prevents them from disclosing the information on these lists during or after their employment. All lists should be marked "confidential." Remind departing employees of their obligations and which trade secrets the business wishes to protect.

NON-DISCLOSURE AGREEMENTS

One of the easiest ways to protect intellectual property rights is through the use of a simple written contract known as a nondisclosure agreement (NDA). These agreements prohibit others from using confidential information or trade secrets for their own benefit or divulging that information to someone else. Time limits are typically placed on the confidentiality obligation. Non-disclosure agreements can be used to protect technical information, potential licensee or customer lists, processes, designs, formulations, and similar intangibles. These agreements, when properly enacted, are legally binding under contract law. Such agreements should be used whenever proprietary information is shared.

Bioentrepreneurs should use and NDA whenever you are going to discuss proprietary information to anyone. It is important to protect intellectual property rights so others on not steal your ideas. NDAs define the treatment of the proprietary information under terms that are acceptable to both parties.

Any public disclosure can destroy the ability to file a patent. The United States has a one-year grace period from any such disclosure, but most foreign countries do not provide a grace period. A breached non-disclosure agreement cannot undo a public disclosure, but confers the right to sue the disclosing party under the agreement.

LAB NOTEBOOKS

One key to protecting IP had been to properly maintain lab notebooks. Lab notebooks record ideas, technical data, calculations, suggestions and any information pertinent to establishing a continuing picture of progress, solutions or failures that bioentrepreneurs encounter. They serve as a means of collecting informal notes and maintaining a chronological record of invention activities. When properly maintained, a notebook serves as an official record of an invention to show the development of it from conception through lab experiments leading to prototypes or completed devices.

Lab notebooks are a good practice beyond documenting ideas for patent protection. As the U.S. transitions to a first-to-file patent system, the use of lab notebooks when arguing about who owns a patent will become less significant. Use of provisional patents or other workflows to speed patent filing will become as crucial to owning an invention as meticulous lab notebooks had been in the past. It is now a race to the patent office where you might file an idea more speculatively than before, even if all the details are not yet worked out for an innovation.

IN SUMMARY

- At the heart of creating value from innovation is the IP protection process.
- IP is products of the mind that can be owned, transferred from one entity to another, and used to generate wealth.
- IP is critical to achieve and sustain an organization's competitive advantage.
- IP provides exclusivity and can be transferred for economic benefit.
- IP provides the tools to enhance the value of an idea and secure it as property that can be bought or sold as the case study below illustrates.

FILING PROVISIONAL PATENT APPLICATIONS EARLY & OFTEN: A CASE STUDY

The Leahy-Smith America Invents Act of 2011 changes the current "first-to-invent" system to a "first-to-file" system. When this change becomes effective on March 16, 2013, filing provisional patent applications both early and often has gained importance as you develop your IP strategy. The example below illustrates this point.

An inventor founded and developed a biotechnology company over a period of 16 years. During that time he hired researchers to work on further developing the company's technologies. His goal was to build and grow this new venture for eventual acquisition by another company. His IP strategy focused on employee submission of written invention disclosures and rapidly filing provisional patent applications. This allowed the company to keep its legal costs relatively low (filing a provisional patent application fees were $500 to $1,000), create early priority dates, combine various improvements after determining how different technologies fit together into commercial products, and decide whether to pursue foreign rights. This was a particularly effective IP strategy for this company. However, it would be ineffective if the underlying technology was not valuable or there were no willing buyers.

After a quick review of the technology, the patent attorney filed provisional applications which included several pages of written disclosure including drawings and photos to better describe the invention with the USTPO to:

1. identify any glaring issues,
2. ascertain any issues with enablement,
3. remove any disclosure language that could be harmful to determining obviousness,
4. look for support for drafting potential claims later on, and
5. determine if the invention could be practiced without undue experimentation and if it actually worked as intended.

Inventors notified the patent attorney of any technology development changes to the invention to determine if filing of an additional provisional patent application should be made to reflect the changes. About ten months after the provisional patent application was filed, the inventor reviewed prior art, discussed with the patent attorney whether to bundle related provisional patent applications filed over the past ten months, and whether to proceed with a formal, non-provisional patent application. Costs for filing a formal patent application ranged from $4,000 to $8,000. However, the complexity of a technology or bundling several related provisional patent applications, *i.e.* multiple inventions in a single application, could increase patent application costs.

After utilizing this "file early and often" strategy for approximately sixteen years, the inventor sold his company for approximately $2 billion dollars based on the patent portfolio he had built.

Protecting your intellectual property can take time, money, and considerable energy—not to mention frequent trips to an attorney's office. In most instances, the cost in both time and energy is worth the protection gained. Developing a strategy with the right attorney or firm will be crucial to achieving the best protection possible within your budgetary constraints.

The difference between the life and death of a new opportunity may hinge on only one or two recognizable ideas or a clever logo or product design. Patents provide an exclusive ability to benefit from an invention. Trademarks, on the other hand, can be a powerful marketing tool for distinguishing the owner's goods or services from those of competitors. They add value to a company. Patents must be filed with the U.S. Patent and Trademark Office to obtain protection and this is a good idea for trademarks also. Copyrights originate automatically upon production of the original, but should be registered if infringement is likely or suspected. Trade secrets also are created with the original, proprietary work and should be used whenever possible due to the ease by which they can be created.

It is important to understand the value of IP and take steps to protect it. IP creates value for the bioentrepreneur and creates competitive advantages over goods and services already on the market.

ACKNOWLEDGMENTS:

Thomas D. Franklin, a patent attorney with the Denver office of Kilpatrick Townsend & Stockton LLP, a law firm specializing in intellectual property and a wide variety of other legal services (*www.kilpatricktownsend.com*), reviewed the material in this chapter for accuracy.

CHAPTER 8:

FDA Approval and Reimbursement

THE FOOD AND Drug Administration (FDA), an agency of the Department of Health and Human Services (HHS), is responsible for overseeing the approval, manufacturing, and marketing of food, drugs, medical devices, vaccines, biologics, animal and veterinary products, cosmetics, radiation-emitting products, and tobacco products. New drugs must be approved by the Center for Drug Evaluation and Research (CDER), medical devices are overseen by the Center for Devices and Radiologic Health (CDRH) and vaccines and tissue products are regulated by the Center for Biologics Evaluation and Research (CBER). This chapter provides a brief overview of the FDA approval process and receiving third party reimbursement for your new life science idea.

CLINICAL TRIALS

Before FDA approval is granted, clinical trials are usually required to provide the FDA approval bodies validated data to make approval decisions. Clinical trials are biomedical research studies conducted on humans that follow a pre-determined protocol. For new drugs, devices, diagnostics or biologics, they are intended to demonstrate safety, and the data derived from the study is usually used to support an FDA application to manufacture and market a new drug or device.[1]

1 http://www.clinicaltrials.gov

Clinical trials are expensive to design and complete. While the specific details of how to run a clinical trial are beyond the scope of this book, you should be familiar with some basic terminology and concepts. The outcomes of clinical trials are major value inflection points for new biomedical companies. The results of the trials can greatly impact the ability of a startup company to move ahead as a going concern or impact the stock price of an established company trying to get a new drug or device to market. The FDA has defined good clinical practice (GCP) and the elements of human subject protection (HSP) necessary for clinical trials (see http://www.fda.gov/ScienceResearch/SpecialTopics/RunningClinicalTrials/default.htmlfor more information).

Before beginning, a clinical trial must be reviewed and approved by an institutional review board (IRB). An IRB, also known as an independent ethics committee (IEC) or ethical review board (ERB), is a committee that has been formally designated to approve, monitor, and review biomedical and behavioral research involving humans with the aim to protect the rights and welfare of the research subjects. In the United States, the FDA and Department of Health and Human Services regulations have empowered IRBs to approve, require modifications in planned research prior to approval, or disapprove research. An IRB performs critical, independent oversight functions for research conducted on human subjects that are scientific, ethical, and regulatory. Following the completion of clinical testing, every new drug must be approved by the FDA before it can be manufactured and sold.

THE FDA APPROVAL PROCESS

Prior to administering a new drug, device, or biologic to humans, the FDA must find the new product to be safe and effective for humans. The proponent must therefore develop clinical data. The first step is for FDA to approve the product as an investigational new drug (IND) or an investigational device. For a medical device, an investigational device exemption (IDE) allows the investigational device to be used in a clinical study in order to collect safety and effectiveness data required to support a Premarket Approval (PMA) application or, occasionally, a Premarket Notification [510(k)] submission to FDA. The FDA approval process varies for drugs, devices, biologics and combination products. Each of these processes is briefly described below.

DRUGS

Drug research, development, and commercialization go through several stages including clinical trials before receiving FDA approval for marketing and selling. Traditionally, clinical trials for new drugs fall into six phases. Progression to the next phase depends on successful completion of the preceding phases.

Preclinical Studies

This phase involves test tube or laboratory studies (in vitro) and trials on animal populations (in vivo) to gather preliminary information on the efficacy, toxicity, and how the body processes the drug which leads to a decision on proceeding onto human-based clinical trials. Once a company has completed preclinical testing in animals, the IND application must be submitted to the FDA for approval to conduct testing in humans.

There are three IND types:

- Investigator IND is submitted by a physician who initiates, conducts, and directs the study.
- Emergency Use IND authorizes an experimental drug for use of in an emergency situation that does not allow time for submission of an IND.
- Treatment IND is for experimental drugs that have shown promise in clinical testing for serious or immediately life-threatening conditions. In this case, the FDA reviews the application while the final clinical work is conducted.

Phase 0 Trials

This phase involves exploratory trials on a 10 to 15 humans conducted under the terms of the FDA Guidance on Exploratory Investigational Studies. During this phase, sub-therapeutic doses of the drug, too low for any therapeutic effect, are administered to determine whether how the body processes the drug (pharmacokinetics) and how the drug affects the body (pharmacodynamics) are as suggested by the pre-clinical studies.

Phase I Trials

This is the first trial on a larger number of human subjects, typically 20 to 100 volunteers. The subjects are studied in an inpatient clinic where they are observed fulltime by clinic personnel. The trial is designed to evaluate the safety, tolerability, pharmacokinetics, and pharmacodynamics of the study drug. This trial also attempts to determine the drug dose necessary to produce the desired therapeutic effect.

Phase II Trials

Several hundred human subjects may be included in this phase of trials to continue the safety assessments started in Phase 11 and to determine how well the drug works. This is the phase which most drugs fail if they are going to fail at all. The trials may reveal unacceptable toxic effects or poor efficacy. This phase is sometimes separated in to IIA, dosing requirements, and IIB, efficacy components.

Phase III Trials

In this phase, random, controlled multi-center trials are conducted on groups of patients that may vary in number from several hundred to several thousand depending on the condition or disease that is targeted. The purpose is to make a definitive evaluation of the drug's effectiveness contrasted with the best currently available therapy for the condition or disease. These trials are the most expensive, challenging, and time-consuming to design and conduct. Possible additional purposes of these trials are to demonstrate the drug's efficacy for conditions or diseases other than those originally targeted (label expansion), to gather more extensive safety data, or to back up specific marketing claims the drug's marketer intends to make. Typically, two successful Phase II trials are completed before their results are gathered into a single volume of background material, *i.e.*, trial methods and results, formulation details, and manufacturing processes, for submission to the FDA. If FDA approval is obtained, the drug can be marketed and sold.

Phase IV Trials

Phase IV trials are called Post-Marketing Surveillance Trials and are carried out after a drugs has received FDA approval and is available in the marketplace. These trials monitor the product's safety including potential long-term risks, optimal use, and provision of technical support provided. These studies may be requested by the FDA or performed by the venture to identify new markets.

Following the successful completion of the clinical trials, a New Drug Application (NDA) which includes the results of the clinical trials of the IND must be submitted to the FDA for final approval of a new drug for commercialization in the U.S.

MEDICAL DEVICES

Medical devices play a critical role in health care. They range from simple tools (*i.e.* tongue depressors and bandages), to complex or life-saving equipment (*i.e.* pacemakers, magnetic resonance imaging machines, and heart–lung machines). A medical device is anything intended to diagnose, cure, mitigate, treat, or prevent a disease or condition that affects the structure or function of the body, does not achieve its intended use through a chemical reaction, and is not metabolized to achieve its affect.

The FDA, through the Federal Food, Drug, and Cosmetic Act (FD&C Act), requires that before a device can be marketed, a "reasonable assurance of safety and effectiveness" must be proved. The FDA categorizes medical devices into three classes:

- Class I – low risk (*i.e.* tongue dispensers or elastic bandages),
- Class II – mild to moderate risk (*i.e.* catheters or hearing aids), and
- Class III – high risk (*i.e.* automatic external defibrillators or pacemakers).

The FDA's General Controls apply to all medical devices while additional specific controls apply to those in Classes II and III. Most devices deemed to have a moderate risk to patients in Class II generally cannot go on the market until they are approved through the FDA's Premarket Notification 510(k) process. Most devices in Class III must be approved through the FDA's Premarket Approval (PMA) pathway.

PREMARKET NOTIFICATION 510(K)

The 510(k) process is a notification process. The 510(k) petition must be submitted to the FDA before marketing commences, and it requires demonstration that a device is substantially equivalent to a device that is already legally marketed within the US. As an intention to streamline the approval process, most Class I devices and some Class II devices are 510(k) exempt (do not require premarket notification). All firms manufacturing medical devices, whether the device is exempt or nonexempt must be register general information with the FDA using its Unified Registration and Listing System (FURLS) and complete an annual registration with the FDA's Device Registration and Listing Module (DRLM).

PREMARKET APPROVAL (PMA)

Class III devices require a PMA to evaluate for safety and effectiveness. To receive a PMA, the device must be proven safe and effective through validated scientific data, *i.e.*, through a clinical trial. In addition, some devices that are found not to fit into Classes I or II require PMA. If a device considered to have high risk and is to be used in a clinical trial, an Investigational Device Exemption (IDE) is needed to conduct the trial. An Institutional Review Board (IRB) must approve the clinical study before it begins. Clinical studies using these high-risk devices must be approved by both the FDA and an IRB beforehand.

BIOLOGICS

Biologics include vaccines, blood or blood components, allergenics, somatic cells, gene and cellular therapies, or tissues used to treat disease. By definition, biologics must be created by a biologic process not chemically manufactured. They are comprised of cells; tissues; or sugars, proteins, nucleic acids, or a combination thereof. Biologics are a hot research topic in the life science industry today due to gene-based therapies and stem cells. Biologics are approved in a manner similar to other drugs.

The term biosimilar refers to products that are marketed after expiration of patents, which are claimed to have similar properties to existing biologic products. Due to the complexity of biologics, a product can only be made that is similar, but not identical. Since biotechnology drugs are only recently coming off patent, the biotechnology industry and the FDA have only just begun to address the issues concerning the sale and manufacture of biosimilars.

Congress passed the Biologics Price Competition and Innovation Act, known as the Biosimilars Act, as part of the Patient Protection and Affordable Care Act of 2010. The Biosimilars Act seeks to streamline the development and FDA approval of biosimilar versions of already-marketed biologic drug products, like generic drugs and drugs similar to biotech drugs, with the goal of lowering prices through increased competition.

The Biosimilars Act is one of the most complex pieces of FDA-related legislation in many years, with enormous financial and public health consequences riding on its implementation by FDA and the courts. The Act raises many difficult questions, and biotechnology manufacturers, potential biosimilar applicants, and even the FDA are still struggling to understand the full range of implications of the new law. While one side argues the uniqueness of a drug made using a biologic process, those interested in making biosimilars point to the similarities in effectiveness and potential cost savings.

For a more extensive analysis of the law and its implications, check http://www.wileyrein.com/resources/documents/BNA_Czaban_May2010.pdf

COMBINATION PRODUCTS

Combination products are defined in 21 CFR 3.2(e). A combination product includes:

1. A product comprised of two or more regulated components, (drug/device, biologic/device, drug/biologic, or drug/device/biologic), that are physically, chemically, or otherwise combined or mixed and produced as a single entity;
2. Two or more separate products packaged together in a single package or as a unit and comprised of drug and device products, device and biological products, or biological and drug products;
3. A drug, device, or biological product packaged separately that according to its investigational plan or proposed labeling is intended for use only with an approved individually specified drug, device, or biological product where both are required to achieve the intended use, indication, or effect and where upon approval of the proposed product the labeling of the approved product would need to be changed, *e.g.*, to reflect a change in intended use, dosage form, strength, route of administration, or significant change in dose; or
4. Any investigational drug, device, or biological product packaged separately that according to its proposed labeling is for use only with another individually specified investigational drug, device, or biological product where both are required to achieve the intended use, indication, or effect.[2]

Perhaps the best know combination product is the drug eluting stent, used to treat patients with heart disease. Part drug and part device, the combination product requires a special approval pathway from the FDA.

REIMBURSEMENT

Getting someone to pay for your product or service can make or break your idea and should be incorporated into your planning in

2 http://www.fda.gov/CombinationProducts/AboutCombinationProducts/ucm118332.htm

the very early stages of new product development. Unlike many other countries, the United States has a decentralized system for paying for drugs and devices with many different players. Getting payment can therefore be a challenge.

In general, the reimbursement process involves three stages:

1. CPT code assignment

   CPT (Current Procedural Terminology) codes are numbers assigned to every task and service a medical practitioner may provide to a patient including medical, surgical and diagnostic services. They are then used by insurers to determine the amount of reimbursement that a practitioner will receive by an insurer. Since all players use the same codes to mean the same thing, CPT codes ensure uniformity. For a description of the CPT code process, check http://www.ama-assn.org/ama/pub/physician-resources/solutions-managing-your-practice/coding-billing-insurance/cpt/applying-cpt-codes.shtml. A recent development has been the adoption of a new CPT coding system, CPT-10, that includes many more codes and sub-descriptions of its predecessor, CPT-9.

2. Getting third party payers

   Third party payers are the health insurance companies and government payers, such as the Center for Medicare and Medicaid Services[3] to include your drug or device as a covered benefit and agree to issue payment to providers.

3. Price

   The amount of reimbursement for the use of a device or drug depends on several market, macroeconomic and health policy decisions. Medicare reimbursement is the typical standard with other private insurance

3 http://www.cms.gov

companies following suit.

Validated clinical evidence and outcomes are the most important things to offer payers and the body of data used to obtain FDA's approval or clearance will not suffice most of the time. Avoiding a surprise reimbursement story includes determining the value proposition of your customer, which includes both the insurer and the provider, both today and in times to come.[4]

SUMMARY

- Receiving FDA approval for a new life science products or services is a long, arduous, and expensive process.
- Many of the most promising innovations fail because they are not approved by the FDA.
- The Food and Drug Administration has outlined specific approval processes for:
  - drugs,
  - medical devices,
  - biologics, and
  - combination products.
- The US has a decentralized system for reimbursements with many different players including government (*i.e.* Medicare and Medicaid), private insurance companies, and individual patients.

ACKNOWLEDGMENTS:

Jason W. Sapsin, an attorney with Polsinelli Shughart's Life Sciences group, leading its FDA practice (*www.polsinelli.com*), reviewed the material in this chapter for accuracy.

4 http://www.medicaldevicedaily.com/servlet/com.accumedia.web.Dispatcher?next=bioWorldHeadlines_article&forceid=73525

Chapter 9:

Developing your Strategy and Business Model

WHAT IS A BUSINESS MODEL?

A BUSINESS MODEL is a description of a business' operations of including the components of the business, the functions of the business, the business' relationships to its suppliers, and the revenues and expenses the business generates. In short, the business model describes how the company will make money in a defined time period. Business models focuses on value creation by representing how the venture will or intends to make money.

A business model also describes the processes a business uses to allocate and distribute resources to make a profit. Some business models are quite simple, *e.g.*, a business produces a product and sells it directly to a customer. However, most business models are a bit more complicated, especially in the healthcare industry where the actual customer is not usually the patient, a third party (*e.g.* insurance company) usually pays for the product or service, and a service provider usually influences the purchasing decision.

Business models can be very innovative and provide companies a competitive advantage. Dell, for example, became famous for its unique manufacturing process that assembled computers only when customers ordered them thereby considerably reducing costs. The business model for a national telemedicine company is much different from a traditional face-to-face medical office. The pharmaceutical industry business model has changed from a vertically integrated business, one that included research, drug discovery, development and sales, and marketing designed to discover and sell blockbuster drugs (*i.e.* those that generate over $1 billion in sales) to a model that outsources drug discovery and development to strategic partners and focuses more on niche products resulting from recent discoveries in genomics and personalized medicine.

KEY COMPONENTS OF A BUSINESS MODEL

A business model should have six integrated parts:[1]

- Customers,
- Solutions,
- Processes,
- Resources,
- Stakeholders, and
- Cash flow.

Your business model should address each of the components.

CUSTOMERS

Who is the customer, and what are their needs and options? The customer component identifies a distinct group with urgent needs that are not being fully met by currently available solutions. The healthcare and biomedical industry is unique in that there are several potential customers including patients, healthcare providers, payers, and service providers. Each occupies a separate place in the value chain (*i.e.* how each adds value to the ultimate product) and act as purchasers, influencers, or users. Consequently, you need to understand who will adopt (doctors), pay for (patient or insurance company) and who will use your products and services and how (patients).

For example, Cartilaze[2] is a company developing a non-surgical

1 http://www.bluemininegroup.com/articles/BMG_What_is_a_business_model_12-02-09.pdf

2 http://www.cartilaze.com

treatment for protruding ears. The target customers are plastic surgeons who perform cosmetic surgery for lop ears. Some insurance companies might pay for the procedure, but like most cosmetic surgery, the patient will usually pay out of pocket. However, the ultimate user is the patient or the patient's parents since most patients are between five and ten years old.

SOLUTIONS

What solution will deliver greatest value to customers and to your company? The objective is to define the products, services, and a customer experience that will deliver a total set of benefits to meet all of the customer's key core and buying needs better than all their other options, at a price they are able and willing to pay. At the same time, your solution needs to create as much value as possible for your company, ideally through a unique formula that allows you to charge a price well above the solution's costs to earn a profit and makes it difficult or impossible for competitors to imitate.

For example, Western States Biopharma developed a new treatment for autoimmune diseases like rheumatoid arthritis (RA). Their solution involves creating drugs that specifically target the defective immune system in patients with RA that results in fewer side effects and better pain relief.

PROCESSES

How can we best deliver this solution to these customers? The processes component is to figure out the optimal set of activities to win customers and deliver the solution to them at the right time, the best quality, and at minimal cost. There are three main sets of processes critical to any business model:

- product development (how to define and design solutions);
- marketing (how to attract and win customers); and
- delivery (how to produce, deliver and support the solutions).

For example, iTriage[3] is a mobile health application that tells patients where to go for care, the location of the office or emergency room and real time waiting time at each facility by using their smart phones or computer. Hospitals and urgent care clinics as well as advertisers pay a fee to be listed on the site.

RESOURCES

What resources do you need to successful sells the products and services? Determine the resources you need in order to develop, market, and deliver your product/service to customers. Physical resources refer to tangible things, such as fixed assets (*i.e.*, facilities, plants, equipment, and tools) and raw material inputs. Human resources encompass capabilities (*i.e.*, knowledge and skills) and connections (*i.e.*, relationships and reputations). Financial resources comprise cash in hand and cash owed to you (*i.e.*, receivables), and the debt or equity used to fund the business.

For example, Medvoy[4] is a global doctor to doctor integrated referral platform for those health care providers interested in extending the reach of their practices to international patients seeking global care. There are no physical assets, since the business is entirely web-based and requires minimal human resources and capital. Medvoy charges users for lead generation and realizes income from advertisers and partners such as medical facilitators and other travel intermediaries.

STAKEHOLDERS

Who is the best team to make this happen? The stakeholder component identifies the right set of owners, employees, and partners to contribute resources, so as to maximize the value for each of them and for the business as a whole.

For example, Cerescan Imaging[5] is a function brain imaging company that provides diagnostic imaging services to patients with degenerative and traumatic brain diseases and neuropsychiatric conditions. Multiple stakeholders, including strategic industry part-

---

3 http://www.healthagen.com

4 http://www.medvoy.com

5 http://www.cerescanimaging.com

ners, investors, providers with an interest in neuropsychiatric conditions, and many others have joined forces to create and finance the company.

CASH FLOW

How will you make money? The cash flow component is to determine the right business model to achieve sustainable, positive cash flow. This requires you to understand the cash impact of all of the other five components. Through a carefully constructed cash flow model, you can explore different options in each of the business model components and assess their economic impact. Over time and through continual iteration and refinement, you can improve the economics of your business model.

For example, Health nation[6] connects patients to doctors via information and communications technologies. To access the services, patients pay a monthly fee and can call a doctor at any time, day or night.

FAVORABLE BUSINESS MODELS

The perfect business model would be one where there is no competition, there are barriers to entry that are impossible to overcome (*i.e.* a capital intensive company granted a monopoly), the suppliers and buyers are at the mercy of the pricing structure of the company (*i.e.* a single seller free to set prices), there are no threats of a substitute (*i.e.* no new technologies likely to replace the product), customers pay immediately in cash, there are no regulatory hurdles, and the company can exit or harvest at any time. Unfortunately, no such model exists. Rather, each company has a model that has strengths and weaknesses that expose it to threats.

Favorable business models are those that have low costs of doing business such as general and administrative costs and sales and marketing costs. Internet-based models, like 2nd md[7], tend to have higher margins. The model involves no cost of goods sold or inventory and connects patients to doctors who are on call and ready to

6 http://www.healthnation.net
7 http://www.2nd.md

assist patients.

The growth of the Internet and social media sites such as Facebook and Twitter have created several new business model opportunities. Many elements of traditional business models are still relevant for online business models.[8] Online business models require the following.

- A killer application which fulfills the target market's needs unlike any other. If possible it is twice as good at half the price of current offerings (*e.g.* there are now over 18,000 iPhone apps for medicine).
- The price does not matter because the target market clearly sees the value and has the means to buy the product without hesitation.
- The price is based on what the market is prepared to pay, not as a set percentage markup on costs (*e.g.* charging for a service based on the reputation of the provider).
- Customers have been waiting a long time for someone to finally offer this product. Therefore, there is no need to educate the market (*e.g.* online scheduling for a doctor's visit).
- There are plenty of opportunities for repeat business, or there is a steady stream of new customers. The product is not a fad or dependent on a technology that will soon be superseded.
- Once purchased, your customers cannot imagine life without your product (*e.g.* a cell phone).
- Significant competitive edge that no other business has access to (*e.g.* stem cell treatments for chronic heart failure).
- A monopoly in a target market that is completely supported by your customers (*e.g.* the only medical device that does what it does).

---

8 https://www.quora.com/Business-Models/What-is-the-perfect-business-model

BIOPHARMACEUTICAL AND MEDTECH BUSINESS MODELS

A striking example of how business models in life science are changing is why many biopharmaceutical companies are moving away from a focus on blockbuster drugs to a focus on core competencies. This is because the annual growth in the global pharmaceutical market is expected to fall from nine percent in 2000 to three percent in 2013[9]. Instead of attempting to create one drug for a mass market, companies are now shifting their strategies to personalized medicine and population disease control. Global giant Astra-Zeneca, for example, has considerably reduced their capacity around the world and has focused on small molecules and biologics, their core strength.

Dimitris Dogramatzis has outlined several biopharmaceutical models.[10] The "blockbuster model" or fully integrated pharmaceutical company model is defined as a vertically integrated company producing a product with annual sales of at least $1 billion or $5 billion in five years post-launch, *e.g.* Prozac and Lipitor. It has been the predominant model until recently. Threats to the model include a dwindling R&D pipeline of new drugs, expiring patents on existing drugs, dropping reimbursements, and the emergence of competitive generics with a resultant fragmentation of drug development responsibilities.

The "orphan drug model" is designed to encourage development of drugs for diseases that affect less than 200,000 people at any given time. There are about 7,000 such diseases. The "virtually integrated model" compartmentalizes the value chain and delegates responsibility for each stage to subcontractors with expertise anywhere in the world. As a result, contract research organizations (CRO), contract manufacturing organizations (CMO), and contract sales organizations (CSO) have grown exponentially.

"Drug development companies" specialize in initial stage research and discovery of products or tools with the expectation that they will hand off the results to later stage companies for product development and commercialization. Such companies might specialize in a given area like pharmacogenomics, molecular diagnostics,

9 http://imshealth.com/portal/site/imshealth/menuitem.a46c6d4df3db4b3d88f611019418c22a/?vgnextoid=500e8fabedf24210VgnVCM100000ed152ca2RCRD

10 Dogramatzis, D. (2010) "Healthcare biotechnology: A practical guide." *CRC Press*

or personalized medicine. Companies using the "biosimilars, follow up or generics model" make and sell drugs equivalent to others that no longer have patent protection. These models create unique challenges for clusters and countries around the world, like Estonia and other Central and Eastern European countries.[11]

While biopharmaceutical companies spent a reported $45.8 billion in drug research in 2009, they spent twice that much on sales and marketing, a business model that is increasingly suspect.[12] The result has been a fewer new drugs in the pipeline. To address that, the US government has created the National Center for Advancing Translational Research, which will provide the infrastructure and services designed to attract drug company investment in drug candidates with promise. The Center plans to open in the fall of 2011 and will present bioentrepreneurs with new opportunities to get their ideas to market faster.

Likewise, the medical device/medtech industry is facing significant challenges due to a shift to outcomes based approval and reimbursement. They are sustaining innovation, delivering value and outcomes, and fueling growth. These challenges have forced fundamental shifts in the medtech business model and how they interact with customers, partners, collaborators, payers, and regulators.

The Healthcare Market Research Publisher reports that the once conservative medical device industry is outsourcing at an increasing rate, and the new tax, which could put pressure on margins, will be one more factor in growing the $60 billion market for outsourcing medical devices and device parts. In order to partially subsidize healthcare financing and reduce the deficit strain, the Patient Protection and Affordable Care Act instituted a 2.3 percent excise tax on "taxable medical device" sales beginning in January 2013. The tax applies to medical devices intended for human use but exempts eyeglasses, contact lenses, and hearing aids, as well as devices that are "generally purchased by the public for retail or individual use," as determined by the Secretary of the Treasury. Two measures recently

11 Suurna, M. (2011) "The developments in business models in the central and eastern european countries: The example of Estonia" *Journal of Commercial Biotechnology* 17, 84–108.

12 http://www.nytimes.com/2011/01/23/health/policy/23drug.html

were introduced in the House and Senate to repeal the tax.[13]

SUMMARY

- A business model describes how you will do business and make a profit.
- Construct and continually modify your business model to conform to rapidly changing environmental and market forces.
- Innovative business models for traditional products and services can create enormous value.

13 http://www.ey.com/Publication/vwLUAssets/Pulse_Industry_Medical_Technology_Report-OCT-2010/$FILE/PulseIndustryMedicalTechnologyReport%20OCT%202010.pdf

Section III
Execution & Implementation

Chapter 10:

Building the Management Team and Infrastructure

Rarely will you have loads of experience and expertise in every aspect of business. Fortunately, you can build a team that fills in your gaps. For those starting a new venture, attracting, hiring, and retaining a highly talented management team that has prior experience with starting a new business or bringing a new product to market is the ideal strategy. For those intrapreneurs or corporate innovators, assembling a strong innovation team to help with the development and commercialization of your invention is also critical. Building a well balanced team with varying expertise is an equally important strategy. Regardless of starting a new enterprise or innovating within an organization, the information in this chapter about the role of the management team and infrastructure is important to learn and understand since the people involved in commercializing an invention can make or break it.

Experience counts even more than having an advanced degree. For example, management teams that have been successful with new startups in the past have an easier time raising outside capital because they better understand what is required, what investors look for, and how to play the financing game. Experience will help you navigate difficult and stressful new startup challenges and help the team foresee potential problems.

Other key factors to consider before assembling a strong management team is evaluating each person's relationships with key customers, professional and industry networks, as well as their experience within your domain — your particular idea. Also important are high energy levels, a strong commitment; a risk profile that matches that of the founders, being adaptable, being a good listener, and recognizing knowledge gaps. Once the team is in place, continuing to build the team is an ongoing process as well as focusing on hiring more qualified staff than the original management team.

Investors invest in new businesses because of their confidence in the abilities, track records, and commitment of the management team. Top management teams are a leading key factor investors look for when financing a company. Why? Predicting which ideas will takeoff is extremely difficult. There's only so much due diligence that can be done on new markets and uncharted territory in the business world. Investors, thus, invest in the people as well as the idea. The idea would be nothing without the people to make it a reality and vice versa. Management team members directly affect the execution of an idea, where it can go, and how far it can go.

This does not mean that a stellar management team guarantees success or that an idea without good management cannot succeed. In today's highly competitive business environment, there is no room for any weaknesses on the team. In addition, the team is what enables you to execute your business model. Strong management team members will pinpoint any weakness and then take steps to address them.

One of the classic traps entrepreneurs typically fall into is placing relatives and friends in key positions. Working with someone who is not highly skilled to do the job and who may jeopardize your success could be disastrous. Inept relatives, friends, and coworkers may quickly lead your new venture into the ground.

This phenomenon is so well recognized that professional investors become immediately suspicious of a feasibility plan or a business plan with a management team made up of people without the necessary skills and expertise to launch the venture or commercialize the idea. You must be prepared to defend the selection of the people who are to play key roles and the value their experience offers.

Ideally, there are four primary ways to justify someone for a position on the management team:

- The person has the track record to prove his/her talents.
- The person has the experience, training, and skills to do the job.
- He/she has money to invest as well as financial expertise.
- They have the connections to help get to the next critical success milestone.

THE TEAM

Depending upon the nature of the enterprise, there are several common roles essential to most new startups. While each of these roles is discussed separately, often two or more roles may be filled by the same person. Commonly, the founder or entrepreneur will assume many of these roles until there is enough revenue and need to justify hiring specialists. For example, the marketer and the sales manager may be the same person. Think about the role(s) for which you are best suited.

DIRECTOR OF R&D — THE CREATOR

A physician, clinician, inventor, or engineer; sometimes a researcher; and occasionally a market-savvy businessperson is usually the venture's creator and quite possibly its founder. This person creates the value sold to the customer. Even though this might seem like the most important position on the team, it is not. While the creators are important, all too often they grossly overestimate their contribution to the enterprise.

Inventors, engineers, doctors and researchers often pose difficult obstacles either by insisting on a bigger piece of the pie than they deserve or thinking that the CEO, marketing director, and sales manager roles are less significant and can be easily filled. One disaster often encountered is when the creator tries to play the roles of the chief executive officer, chief operating officer, and the marketing director combined. A few manage to pull it off, but not many. Creators usually excel when they do precisely what they like to do and what they do best — create

A temperamental creator often scares professional investors. They wonder if the management team is prepared to continue on if the creator leaves, or if the creator holds all the keys to the show. The team must be certain it is not placing itself at the mercy of the technical side of the enterprise. The team should have a contingency plan ready if the creator leaves. It should be able to identify to investors some people who can replace him/her. If not, the team should be able to show them why the creator will not abandon the enterprise. Business skills take many years of education and experience to learn. A new venture neither has the time nor the resources to teach the creator these skills.

CHIEF EXECUTIVE OFFICER (CEO) — THE DRIVING FORCE

Every management team needs someone who can create and communicate the vision of the business. The key to a venture's success is the entrepreneurial leadership — the person who provides, vision, direction, and inspiration with palpable energy and passion. Without an able driving force, the venture will not amount to much. Thousands of promising ideas have failed because there was no driving force.

For example, after four years and $4 million, the inventor of a new medical technology with great potential was still muddling around in the laboratory. There was no driving force in the picture. Nothing was going to happen. Nobody knew what to do to make same thing happen, let alone how to do it. Experience has shown that most successful entrepreneurs are driving forces, not necessarily creators.

CHIEF OPERATING OFFICER — THE ORGANIZER

Every management team needs someone to make certain that day-to-day operations flow smoothly and economically. They make certain that the necessary work is done on time and properly. The venture must be structured and able to function on a daily basis. All too often organizational and operational talents are precisely the ones lacking from the typical driving force or innovator who, more often than not, hates details. The organizer must relish details.

It is easy for the driving force to hold the execution of details in contempt. However, the execution of those details is what distinguishes the highly successful enterprise from the short-lived venture. Steve Jobs, cofounder of Apple, attributes the company's success to his fortunate hiring of an experienced organizer Mike Markula. Markula was one of the early employees of Intel and was heavily involved in growing Intel into what it is today. His understanding of the chips industry and of building an organization was important to enhance the technology that led to the development and success of Apple's Macintosh product line.

MARKETING DIRECTOR — THE MARKETER

Investors demand marketing savvy. Without marketing capability, success is unlikely. One of the key elements to assess in evaluating a new venture is its marketing plan and team. Investors try to determine if the marketing director can plan, price, package, promote, and place the product successfully in addition to selling it. All sorts of businesses create products with great adeptness. However, marketing and selling the offerings can be another matter.

SALES MANAGER — THE SALES PRODUCER

The sales manager works hand-in-hand with the marketing director. The sales manager plans and organizes the sales team to sell to the target markets selected by the marketers. Sales obviously drive the success of a venture, and this function should be performed by people who understand the industry, recognize benefits motivate the customer, and have the ability to persuade customers to purchase the product/service.

CHIEF FINANCIAL OFFICER — THE WATCH DOG

Outside investors and funders are insistent that the CFO have an excellent understanding of a start-up venture's needs in terms of cash flow and satisfactory internal organizational controls. Depending on the stage of development, a business might need a book keeper, a controller, and/or a CFO. The CFO is the person who works closely with the CEO in establishing the future of the new venture particularly when it comes to raising money or entering public markets.

PRODUCTION DIRECTOR — THE MAKER

Is the new venture going to manufacture its own products and services? Excellent, experienced production professionals are not easy to find, and investors are not going to put money into a manufacturing venture without a top production manager on the team. This is one reason to subcontract the manufacturing so both potential investors and customers are assured that the manufacturing processes are in competent hands.

SHAPING THE TEAM TO THE MISSION

To a great degree, the people hired for each position and the importance of each position depend a great deal on the venture's mission and strategic goals. If the venture is involved in an industry that requires great marketing skills to be successful, then marketing and sales positions must be filled by people who have extensive experience in this industry. Likewise, in a venture that requires highly technical expertise to be successful, then those positions on the management team will be critical.

THE TEAM AND THE BUSINESS PLAN

Organizational leadership and outside investors, if needed, are usually keenly interested in knowing the experience and expertise of the management team. The team's qualifications including their experience should be emphasized in the business plan. Capsule descriptions of the qualifications of the key people should be put into the body of the plan. The details provided by a resume should be put in the appendix. When the venture seeks public money, financial institutions will do a thorough investigation of the team. Team members must be squeaky clean. It is not good to have a team member whose record cannot stand scrutiny.

MOTIVATING THE TEAM

The company will not survive in the long run if the management team is not properly rewarded. As one successful entrepreneur said, "I want my people to get rich because I have set up a system whereby if they get rich, so do I." The wealth generated by the team should be shared among the team members. The only way quality employees can be attracted and retained is to make it worthwhile for them. Fortunately, they generally will not make immediate demands upon the firm's cash flow. They know they will get theirs when they've earned it.

Few topics have been more discussed and written about than employee motivation. Wars are won, teams are victorious, and fortunes made by those leaders who have mastered the art of motivating their people to exert great effort in pursuit of their goals. There seems to be more variation in people's motivation than in their basic aptitudes. Given the proper motivation, skills can often be acquired. Without motivation, little happens.

RECRUITING TOP TALENT

Attracting and hiring top talent is crucial when building the management team. Recruiting the wrong people will directly affect productivity and the bottom line. Recruit from industry experts that you read about in trade publications or meet at industry association meetings, knowledgeable people in the business, infrastructure specialists such as lawyers and accountants who work in the industry, and recruiting firms specializing in the field. These sources are rarely candidates themselves, but often can identify potential candidates or stimulate word of mouth so people self-identify. Running online job postings or help-wanted ads has little chance of finding the right kind of talent.

Once you have identified a few qualified candidates, check professional references by verifying the basic information provided and search for gaps in employment. Such gaps can signal potential problems, such as being fired from a position, not being able to find employment, or even spending time in jail. People who overstate their job qualifications may cause future harm to an organization or perform poorly.

Experts claim that 90 percent of all hiring mistakes can be prevented through proper background checks. Besides, claims of negligent hiring against employers make it essential for a firm to conduct thorough background checks on all potential hires. Studies show that about 40 percent of resumes contain some kind of false or exaggerated information. There are agencies that specialize in doing background checks.

At least two references should be obtained on candidates under consideration. When checking references, former employers should be asked about the applicant's dates of employment, title, duties and functions, stated reason for leaving employment, and whether the former employer would rehire the individual. Also, inquiries should be made about the quality of work, attendance and punctuality, ability to get along with coworkers, and overall job performance. References may/may not provide all of this information because of possible legal liability. In any case, a written record should be kept on all reference checks and the information obtained. A firm must be prepared to back up a hiring decision with written proof that it was based on relevant job information.

If a job applicant does not receive a favorable reference from an employer, the candidate should not automatically be rejected. In this case, several employers can be queried to get a well-rounded view of the applicant's job performance. If possible, each significant employer should be called for references before a final selection is made.

The references should be asked to comment generally about prospective candidates. Additional information will help in making a better hiring decision. A lot of grief can be saved by spending more time choosing the right applicant. Involve others in the interviewing and checking process. Another person may see/hear something that you do not. If other management team members are already hired, get their input. Personnel selection is one of the major decisions early-on and throughout your company's history.

COMMON HIRING MISTAKES

There are some common mistakes that should be avoided when hiring employees or members of advisory and management teams.

- **Inept associates**

  Verify the requisite skills of fellow employees with whom you currently work to ensure they have what you need in the new business. Employing friends and/or family is very risky as their skills may not only be inept, but also you often can begin questioning your own ability to judge other people. It is difficult to admit being wrong about someone you love. Sometimes the management team may be quite able, but the controlling creator proves inept. Thus, the minority associates are posed with the sticky problem of dethroning an incompetent leader.

- **Personality conflicts**

  A business relationship can be as close if not closer than a marital one. When people cannot be open and honest with one another, troubles are likely to arise. The team should be fully comfortable with the person they plan on hiring. Otherwise, the new recruit will not work out. Whatever concerns arise during the interviewing process will be magnified down the line. Good working relationships require honesty, openness, and tolerance and forbearance of others' habits and mannerisms. Mutual respect is necessary. But no matter the precautions, mistakes will be made, and people can change for the worse. After all, no one is perfect. Everyone has flaws as well as strengths.

- **Disagreements over strategy**

  Honest differences of opinion often arise over how a venture should be operated. Typically when the venture is highly profitable, such conflicts are usually minimized, although disagreements do arise over growth strategies. However, if the venture is going through difficult times, disagreements will be plentiful over what should be done about it and who is to blame. Even best friends will have fall-outs over serious disagreements about what to do with a fledgling business. By openly and respectfully discussing such disagreements, they can be resolved as quickly as possible with some consensus; nevertheless, such consensuses not always possible. Differences of opinion do arise which often cause relationships to crack, and you may have to say farewell to some associates.

INFRASTRUCTURE

Only a portion of the work that must be done for a venture to be successful is performed by its own employees. Much work is outsourced to external resources. Management teams often use independent professionals to provide specialized advice. Herein lays a problem: whose advice to seek and whose advice to follow? Rare is the management team that cannot tell terrible tales of difficulties caused by using the wrong people. An incompetent lawyer or accountant can cause irreparable damage. Thus, selecting professionals is of critical importance and you should give these decisions considerable thought and investigation.

The most common independent professionals used are:

- Advisory Boards
- Attorneys
- Accountants
- Bankers
- Consultants
- Manufacturing Contractors
- Distributors
- Board of Directors

ADVISORY BOARDS

This board consists of advisors who act as consultants for the CEO and top managers of the company. The management team should select professionals who understand the new company's market, have expertise in an area that the management team does not, and can open doors. Five to six people with experience and specific expertise in various areas, such as finance, new venture start-ups, marketing, and life science technologies, usually serve on the board and meet four to six times a year. It is recommended that the board be paid for each meeting either with money or some kind of "in-kind" benefit. The advisory board does not have any legal liabilities to the venture and only gives the team advice — it makes no decisions. The team has no obligation to use any of the suggestions from the advisory board.

ATTORNEYS

In this highly complex legal society, finding a competent attorney to provide useful legal advice is critical especially for life science ventures. Attorneys who are hired should have considerable and practical experience in start-up business affairs, in your industry, in intellectual property law, and so on. Select an attorney with the experience and expertise that fit the needs of your venture, and collect all necessary information before meeting with the attorney. You will benefit greatly from doing your homework and being well prepared.

ACCOUNTANTS

There is great virtue in being able to afford the services of large, internationally known accounting firms. They can do many things for business owners that the smaller local companies cannot. They are particularly useful in their worldwide coverage. Whether large or small, an accounting firm that does not provide quick turnarounds on data can be detrimental to the venture. Management teams need to know quickly how they are doing. As when selecting an attorney, also select an accountant with expertise not only with small businesses but also in your industry.

If the management team plans on going public or using bank

money, the team will find it very helpful to have a large accounting firm auditing the books right from the beginning. The Securities and Exchange Committee (SEC) will not accept the work of just any certified public accountant (CPA) for its purposes. Much time and money will be saved the books and filings are done correctly from the start.

BANKERS

Sound banking connections are essential to new companies. A banking relationship should be established early on when founders begin to build their new ventures. Banks can assist in setting up checking accounts, depositing payroll checks, and approving bank-issued credits cards such as a Visa or a MasterCard. Banks can provide capital in form of debt, *i.e.* loans, to companies that have an established track record, a company history, historical financial statements, and a good collateral base. The management team should understand that if good banker relationships are established, the banker will become an important player in the infrastructure and a potential source of financing as the business grows. Bankers have to get to know the management team, build trust, understand the operation, and get a feel for how good they manage. A management team cannot expect to walk into any bank and expect to get money particularly in today's tough lending environment.

CONSULTANTS

Entrepreneurs and new ventures use consultants because they are usually less expensive than employing someone with equivalent expertise. Contracting a consultant particularly pays off when you need a specific problem solved or have a project that needs additional expertise.

There are many skilled consultants who can provide invaluable services such as setting up the books, forming the organizational structure, management, marketing, public relations, human resources, and finance. In no other field are options so abundant, expertise so diverse, and the quality so variable.

MANUFACTURING CONTRACTORS

Most life science ventures do not manufacture or produce their products. Instead, they contract with a manufacturing firm that will manufacture their product to their specifications. There are thousands of firms that exist solely to produce goods for others.

DISTRIBUTORS

A distributor can be many things and take many forms. However, in the end, the distributor is an existing firm that will buy a venture's product and resell it to the intended market. It does just what the name implies; it distributes products for a venture, thus saving much time, money, and anguish. For example, movies get into movie theaters, onto DVDs, and onto iTunes via distributors like Warner Bros. Entertainment that also promote and advertise films. Attaining distribution for a new product is a difficult undertaking, one that often proves to be the entrepreneur's undoing.

BOARD OF DIRECTORS

The board of directors usually is made up of investors and the management team and is created when the venture uses outside financing. The board is responsible for setting policies for the new venture's day-to-day business. Many of their decisions are made to protect the stockholders of the company. The management team does not always get the advice needed from the board of directors because of different agendas, growth versus payback for example. The board has a legal liability from the stockholders if members do not make appropriate decisions when setting the policies for the business. The board should receive payment of some sort to ensure monthly attendance at meetings and involvement in decisions.

IN SUMMARY

- Forming a venture's management team is an art and critically important in getting a new technology to the marketplace.
- Adept selection and use of infrastructure professionals

supplements your venture team's expertise.

- Appropriate professional help is often the difference between success and failure.
- Bankers, lawyers, accounting firms, and advisory boards are critically important and working with them cannot be avoided.
- An effective management team and infrastructure can be the ticket to success, while a poor one may cause the venture to fail.
- Experience shows that the selection of your associates is critically important to success.

Chapter 11:

Biomedical Intrapreneurship: Employee Innovation

Most healthcare professionals work for an employer. Estimates are that more than 60 percent of physicians in the U.S. now work for someone other than themselves, while only about one percent of registered nurses are self-employed.[1] The reasons include lifestyle choices, generational work-life attitudes, the increasing costs and hassles of managing private practices, and hospitals wanting to attract Accountable Care Organization (ACO) dollars.

A byproduct of employing large numbers of healthcare professionals is the opportunity to identify, develop, nurture, and then monetize ideas because the employer, when properly structured, owns the intellectual property (IP) rights to employee ideas, inventions, and discoveries. Consequently, every healthcare provider should create a technology commercialization process, accompanied by IP policies and procedures to identify, vet, maturate, and package their intellectual property assets. This is similar to what technology transfer offices do at most major research universities.

The major barriers to medical staff participation in commercialization efforts are substantial including:

---

1 Kocher, R. and Sahni, N. (2011) "Hospitals' race to employee physicians — The logic behind a money-losing proposition" *The New England Journal of Medicine.* 364:1790-1793.

1. increasing clinical, teaching, and administrative demands on staff time;
2. little value placed on commercial research and development efforts;
3. lack of recognition for commercialization efforts;
4. a gap in understanding the differences between a community-based healthcare and an industrial culture;
5. lack of sufficient financial incentives to motive staff to participate in the process;
6. lack of knowledge, skills, and abilities to participate in more complicated commercialization aspects, such as spinout formation; and
7. weak internal and external networks that leave medical staff unaware of the commercial opportunities available and the accomplishments of their colleagues.

Healthcare organizations should embrace innovation, understand intellectual property ownership issues, and engage in technology commercialization opportunities by encouraging their medical staff to participate in these activities. Healthcare administrators who make innovation a strategic initiative benefit by:

- improving their recruiting, developing, and retaining of clinician talent;
- maximizing the creativity of their workforce;
- better differentiation from competitors; and
- monetizing new ideas resulting in new revenue.

INNOVATION IN ORGANIZATIONS

No organization, especially those in healthcare, can expect to survive, let alone thrive, without capitalizing on their workforce's collective talent and ideas. However, making innovation a strategic initiative relies on an organization's supportive ecosystem, a culture of innovation, staff awareness of commercialization opportunities, removing barriers to participation, providing incentives, and celebrating successes. Each of these core strategies is discussed below.

DEVELOPING A SUPPORTIVE ECOSYSTEM

Healthcare organizations should create an ecosystem that supports innovation and facilitates technology commercialization. It is important that the executive team enthusiastically supports innovation as a strategy and commits resources to ignite and diffuse it throughout the organization. This helps create a strong innovation mindset which will see the stage for all innovation initiatives going forward. The success of an innovation initiative depends on the leadership team embracing it, making it a strategic focus, and committing resources to support it.

Using internal marketing communications channels, websites, and other vehicles, management must inform all staff of the organization's commitment to innovation and commercializing new products and services and implementing process improvements. They should also focus their efforts on those medical staff who are more interested in life science technology commercialization. Executive leadership and administrators should work with clinical champions and respected opinion leaders to motivate others to participate and provide clinical bioentrepreneurs with the connections and support they need to move their ideas forward.

CREATING A CULTURE OF INNOVATION

Healthcare organizations should create a system-wide culture of innovation since culture and innovation play a critical role in their continued success. A culture of innovation also will enable the organization to continually introduce new ways to deliver care as well as develop new business models. Creating a culture of innovation needs to be strategic, *i.e.* the innovation methodology must be a proven and a repeatable process that creates a significant difference in the value it delivers to patients, partners, and the organization. A strategic culture of innovation will generate a robust portfolio of breakthrough business growth opportunities utilizing a two-pronged approach: a disciplined methodology and a creative process.

INNOVATION AWARENESS CAMPAIGN

An innovation awareness campaign aims to build awareness and educate through a marketing plan developed to introduce innovation to internal and external users as well as key stakeholders. Typically, "Innovation Champions" are identified and recruited to help market, socialize, and build awareness about the new innovation initiative encouraging staff to participate thus creating an innovation pipeline.

Short training and education modules should be developed to promote the innovation initiative. These modules should define high potential ideas and explain how to submit ideas, how ideas are vetted, and how they move through the evaluation and incubation system. Sometimes, Ideation Sessions are held to jumpstart the process of focused idea exploration which includes identifying and initially evaluating new ideas from knowledge workers.

REMOVING INNOVATION BARRIERS

Once staff are informed and engaged in the innovation process, barriers to participation should be identified and removed as well as incentives created to participate. Removing barriers involves support from key executive leadership including the appointment of a person to coordinate and be held accountable for the strategic innovative initiative. Identifying organizational innovation barriers is the first step. Once identified, time and effort must be spent developing interventions to reduce and/or remove the barriers. Identified organizational barriers should be monitored and measured to ensure the inventions were and continue to be effective.

PROVIDING INCENTIVES

In addition to establishing favorable revenue distributions from licensing royalties to inventors, other incentives should be created. Examples of commercialization incentives include peer recognition events, publicizing successes both internally and externally, awarding time off to pursue commercialization efforts, developing key performance indicators that recognize entrepreneurial activities, recognizing commercialization efforts on the employee performance evaluations, and innovation and knowledge transfer exchange programs such as internships and sabbaticals.

CELEBRATING SUCCESSES

Celebrating an innovator's successes is important to achieving and sustaining technology commercialization success. By using social media, public relations outlets, internal communications devices, and special events, leadership can highlight even the smallest successes and leverage them to encourage larger participation and accomplishments. Part of building the culture of innovation involves having an ecosystem that champions the accomplishments of life science innovators on a regular basis. Internal and external marketing techniques should be repetitive, redundant (*i.e.* publicizing the same message using different media), and relational (*i.e.* linking one message to another) so that staff members become "owners" and recognize commercialization as a valued pathway to professional and personal success. For example, to publicize a newly licensed invention, a healthcare organization issues a press release which is linked to the commercialization office's website. The announcement is also concurrently posted on the organization's Facebook wall and a Twitter feed is sent to followers.

In summary, filling the innovation funnel with medical staff ideas with commercial potential primarily depends on:

- a growing research or clinical base stimulating discovery,
- policies and structure that stimulates use,
- a culture of innovation that encourages and rewards participation, and
- marketing and public relations mechanisms to celebrate success and spotlight champions.

Each hospital, hospital system, or medical group practice has local and regional assets and constraints. Leadership needs to develop organizational strategies and tactics based on their individual resources, strengths and opportunities. By emulating best practices from other healthcare service organizations, universities worldwide, and research institutions, community-based hospitals and medical group practices can also successfully embrace innovation and stimulate idea and deal flow.

CORPORATE INNOVATION

When most people think of entrepreneurship they think of striking out and creating a new business. Although becoming an entrepreneur is a great path for many individuals, it is not the only way to commercialize a new product or service. Often innovators work within their organizations to commercialize a newly discovered technology—especially when their organization has an entrepreneurial culture and supports innovation as discussed above. This is a great way build one's career as well as to make a difference within an organization. There are benefits to being an entrepreneur or a corporate innovator.

Corporate entrepreneurs, corporate innovators, or intrapreneurs are individuals inside of existing organizations that discover and create new businesses, products, services, and processes that generate value and create new revenue growth. Corporate innovators leverage current organizational competencies or evolve new ones through innovation for the purposes of growth and renewal. New product and service innovation as well as new process and service delivery innovation involves entrepreneurial behavior inside established organizations. Corporate entrepreneurship can be a more formal or quite informal process, but either way, its aim is to commercialize new products, services, or processes which create value and generate new revenue streams. Innovation is a vital strategy for rejuvenating and revitalizing existing organizations.

CORPORATE INNOVATION PROS AND CONS

There are multiple challenges and risks associated with being a corporate innovator which should be weighed and evaluated. Consider both the pros and the cons before deciding to become more entrepreneurial. We believe that the positives far outweigh the negative aspects. The following is a discussion of both the positive and negative aspects of corporate innovation.

Corporate Innovation Pros

There are several advantages of launching new products or ideas under the corporate umbrella.

- **Use of the Corporate Name**

  When launching a new venture, both personal and business credibility of the founder(s) is critical. Lack of credibility is the key reason why many good business concepts fail. The innovator frequently encounters difficulty in convincing potential customers, stakeholder, and/or investors that he or she can perform by delivering the goods/services, is reliable, and has the proper financial backing to support the business venture. The life science professional may have significant technical and professional credentials and an impressive educational background but that does not translate into becoming a successful corporate innovator. Many people do not want to try "experimental" treatments, products, or services. If the treatment comes with a trusted name or the technology is being produced by a well-known and established enterprise, the life science and healthcare markets are more likely to positively react to it.

  A good example of utilizing a well-known corporate name in the consumer products industry is 3M's Post-it™ notes. Art Fry, who developed, manufactured, tested, and marketed Post-it™ notes, used the 3M name and backing to get it to market. Fry may have had a very different experience had he tried to introduce the Post-it™ note on his own and would have probably run out of money in the process since it took approximately seven years to get it to market.

- **Use of Board of Directors**

  In any sized enterprise, a strong board of directors that provides real, strategic input and guidance is invaluable to the organization's executive team. An experienced board brings credibility to an enterprise that often makes fundraising for R&D money easier. In addition to experience and credibility, the board may be connected to individuals who can help obtain the resources required to introduce new products and services. These resources could be customers, suppliers, or even competitors. Having a seasoned board of directors in the life science industry is a great asset to a corporate innovator who is able to access their expertise. It is much more difficult for an independent entrepreneur to recruit an experienced board of directors or advisors and get seasoned mentors who can guide him/her through the turbulent ups and downs of a new life science start-up.

- **Strategic Fit**

  Strategically matching where the organizations headed in three to five years is critical for any new innovative project or technology. Understanding the organization's strategic plan gives corporate innovators the opportunity to align their visions with their organization. A definitive and evident link between the entrepreneurial activity and the organization's strategic direction should be established when exploring new business opportunities. All feasibility and business plans should clearly prove how the new opportunity fits into the organization's strategic plan.

- **Use of Marketing Clout**

  The cost of introducing and marketing a new product/service is most often grossly underestimated. Experience in the industry is the key. Therefore, the existing enterprise can provide the market experience and resources that an entrepreneur would not have on his or her own. This is especially true with marketing life science technologies. Most corporate innovators lack marketing expertise and know-how understanding how to get their product or service to the market. Utilizing the organization's marketing department and existing market connections or patient bases is a great asset. For example, when a life science innovation project fits into established distribution channels of the organization, it makes estimates of the real marketing costs and amount of time required to penetrate the market more reliable.

- **Use of Internal Financing**

  "Money, money, money," is the continual cry of entrepreneurs. They frequently wonder, "Where will I find it? Who will invest in my venture? How much of my company will I be required to give away?" The curse of the entrepreneur is constantly spending time raising funds to support the financial needs of a new venture rather than spending time developing new products or services, devising ways to penetrate the marketplace, and providing excellent customer service. This is particularly a problem for life science start-ups where the up front expenditures are large and the payoff is very far down the line. Getting investors or finding the required amount of funding is always a major problem that can be eased using the financial resources of an existing enterprise. The organization can supply the needed capital or use outside funding sources.

- **Use of Technology**

  Many entrepreneurs spend days, weeks, months, or years struggling to develop the needed technology to make their products work. Sometimes they run out of money to complete the new product development even after they have quit their jobs and taken out second mortgages on their homes.

In the life science industry, getting a new product ready to market usually takes a significant amount of technology maturation. The tools, equipment, and resources required to get a working prototype that can be mass produced are usually quite advanced and very expensive. Without access to these resources, the innovator may not be able to succeed. An additional concern is the availability of the technology. Some ventures require the use of technologies that are not available in an existing enterprise. Gaining access to using the necessary technology will be critical.

- **Use of Top Management as Mentors**

  Having a dedicated and respected upper management mentor can be a huge benefit to the corporate innovator. The board of directors is not the only group in an existing enterprise that can be of benefit to a corporate innovator. The upper management may have as much if not more valuable experience developing new products in the life science market than the board. They can mentor the innovators about next steps to take and can act as a buffer between the board or other risk adverse groups within the organization. If there is no upper management support by a champion or mentor, the chance to successfully develop a new technology using corporate resources is diminished.

- **Use of Corporate Contacts**

  A valuable tool for corporate innovators is having a good network of seasoned executives and people with entrepreneurial expertise to tap for advice. Being able to access corporate contacts that are respected experts and can provide valuable advice is invaluable. Building a strong network of professionals to access for advice is more difficult for the independent entrepreneur. Sometimes the weakest link in an entrepreneur's network is lack of the right contacts who can provide sound advice as well as become supporters and/or customers.

- **Use of the Corporate Infrastructure**

  As mentioned earlier, an entrepreneur needs many different key resources supporting the new venture. This supporting cast is termed the infrastructure. Entrepreneurs must spend time and money finding and hiring competent attorneys, accountants, and scientists with expertise in the proper area of the life science industry. However, when innovating within an existing enterprise, most of the needed professionals in the infrastructure are already in place.

  However, many professionals in the corporate infrastructure do not adequately understand the needs of corporate innovators and new ventures. They do not have experience in the entrepreneurial world and do not realize how important time is when launching a new product/service. Additionally, some enterprises do not have staff familiar with the part of the life science industry in which the new venture plans to operate.

- **Use of Staff Support**

  The typical entrepreneur has low support needs; he or she often acts as the receptionist, marketer, administrative assistant, sales force, and so on. In contrast, employees involved with corporate innovation projects usually have the advantage of using an established support staff. In the life science industry, the people who come up with the idea are sometimes not used to performing all of the necessary tasks and are typically more focused on the technical aspects of their innovation than on the day-to-day execution of the office and business.

Another benefit of working with known staff is understanding their strengths, weaknesses, and work history. It can help you choose an internal management team to initially work with. On the other hand, entrepreneurs are continually faced with the difficult task of hiring the right people to work on a new innovation.

- **Shorter Time to Market**

  Successful entrepreneurs bet whether they will reach cash break even before they go broke. The faster the entrepreneur can get his/her product or service to the marketplace, the better the odds for winning this bet. If corporate innovation projects can utilize many of corporate assets mentioned above, they have a better chance of winning that entrepreneurial bet. Timing and getting new products and services quickly to the marketplace can mean the difference between success and failure. In the life science industry, this can mean the difference between getting a patent and losing out to another organization. Developing the new life science product or service internally enables the corporate innovator to consult with corporate IP counsel and other legal resources to help get the innovation to the market more quickly.

- **A Win for Both the Corporation and Employee**

  A crucial test for guaranteeing the success of entrepreneurial projects is to ensure that both the organization and the innovation team benefit. The first winner must be the organization. What drives a corporation is earning new profits that will ultimately increase the value of its stock. As long as the corporate innovation is meeting bottom-line projections, organizational support should be intact. It is critical that entrepreneurial activities conform to the corporate culture and fit with its mission statement, as well as its present and future strategic plans.

The second winner is the innovation team who is recognized and rewarded for their successes. Many problems occur when management neglects to reward innovative employees for starting, developing, and directing entrepreneurial projects. The organization must ensure that it does not exploit their innovators.

To address reward corporate innovators, many organizations are creating programs to create similar benefits to those that would be received if the entrepreneur left the organization. For example, Bell Atlantic's corporate entrepreneurship program, Champion, provides seed money, mentoring, and entrepreneurial training to employees from any level of the company who propose new products or services. The employee can invest a portion of his or her wages in the project in exchange for interest in the product when it is marketed.

Corporate Innovation Cons

Because innovation projects or new start-ups under the existing corporate umbrella are dependent on the enterprise's resources, the innovation must be both politically and technically correct. Therefore, it is paramount to think about the negative aspects of commercializing the innovation within the scope of an existing enterprise. If the decision is to go forward, the key to its success is to develop contingencies to overcome the negative aspects discussed below.

- **Corporate Culture**

  Terrence E. Deal and Allen A. Kennedy popularized the term "corporate culture" in their book *Corporate Culture: The Rites and Rituals of Corporate Life*. Essentially, this term means "the way we do business around here," and every organization has a corporate culture. Corporate culture encompasses the company's values and standards that are communicated to employees by management style, dress, expectations, and assumptions within the organization. It is the company's personality and credo.

Life science corporate innovators must first realize that the corporate culture is deep-seated and difficult to change. Corporate innovators must work within the corporate cultures and accept the organization's attitudes and philosophies. Sometimes the corporate culture is responsible for the failure of entrepreneurial initiatives. Within the life science industry there may be a lack of understanding of what it takes to get a new technology commercialized. A lack of entrepreneurial understanding combined with a culture that is afraid of change could mean disaster.

- **Hierarchical Structure**

  It has been quite common for organizations to be managed by a highly structured, bureaucratic design containing numerous levels of decision making and complex controls. The typical organizational structure inhibits the innovative process. Organizations, similar to the human body, have systems that react negatively to change and innovation. Multiple levels of management approval can result in serious time lags. Oftentimes obtaining management's blessing at each level in order for a new project to move forward is necessary. In larger organizations, the control systems are usually more rigorous which can cause even more delays.

  Innovation projects and newly discovered technologies require an environment that lends itself to some degree of autonomy, quick decision making and flexibility, and the ability to act independently without obtaining management approval at each step of the way. Chuck Blazevich, Motorola's director of corporate quality for business systems, once said, "There's a strong belief by the leadership of the company that to move ahead we need to empower employees. We don't let layers of management get in the way." This applies to life science innovators working in an organization where managers either look over their shoulder and try to micromanage or are disinterested in the new project. This can be toxic to a corporate innovator. Instead, a more flexible management approach is required which is sometimes lacking in an existing enterprise.

- **Bureaucratic Management Style**
  A bureaucratic management style, the antithesis of an entrepreneurial management style, frequently goes hand-in-hand with the hierarchical structure. Bureaucratic managers tend to block innovation. The way they have learned to manage does not foster and nurture the innovative process. Many times the chance of continuing to work on an innovative project may hinge on proving, in a relatively short period of time, that the project will be successful regardless of the roadblocks encountered. This scenario is particularly difficult in the life science industry, with long time lapses between development, regulatory approval, and positive cash flow or even profit. With upper level managers demanding tangible results on a quarterly basis, it is difficult for bureaucratic managers in an existing enterprise to break out of that focus and see the long term benefits of innovation.

- **Difficulty Managing Mavericks**
  Usually life science innovators have a strong commitment and a strong drive to see their ideas become realities. They start with a vision, develop it, and, against all odds, ensure that it gets implemented. They are indeed action-oriented and are willing to do

whatever it takes to achieve their objectives. In this process, they tend to disregard rules, regulations, and policies.

Consequently, corporate innovators can be a challenge to manage. "No" is a word they frequently disregard. Sometimes they ignore management directives that get in the way of achieving their goals. Day-to-day job duties could be neglected while they are working on a new technology or project. Regardless of their drive and sometimes overzealousness, they should be nurtured and mentored.

- **Risk-Taking is Discouraged**

  An entrepreneur cannot have the fear of failure pursuing new opportunities. Failure is a learning experience and a powerful building block for future success. Innovation is impossible without some degree of risk taking. It can be easy for a manger to become risk averse in a life science industry especially after one or two projects have failed, and they do not want a track record of failure. However, one successful venture can make up for several failures, but that does not always seem obvious in the moment.

  Management tends to collect and overanalyze stockpiles of data before a decision can be made as it tries to avoid risks. However, this "analysis paralysis" stifles the intuitive and creative decision making that is essential for innovation. Lastly, employees may try to avoid risk, since in the organization's environment, their future depends on successful performance in order to move up the corporate ladder. The fear of failure that is prevalent in many organizations in the life science industry is hard to erase.

- **Turf Wars**

  The success of corporate innovation relies on how

well the innovation team can assimilate the corporate support they need to develop the technology. The accounting department, finance department, human resources department, marketing department, and legal department must work well together as a team. Each of these departments has its own culture, agendas, and reward systems, and often delights in beating its peers in the race to the top. Typically these departments have not been required to integrate and work together effectively.

Innovation greatly suffers from corporate turf battles. This is especially true in the life science industry which has many moving parts and all of which requires much cooperation. The innovator needs to be aware that turf battles also occur between colleagues fighting for a raise, promotion, or recognition. This can be a major problem that occurs when innovating within an existing enterprise. For example, someone else in your healthcare organization might have an inside track to people or to money. Or your technology might present a conflict with intellectual property that is being currently developed within the organization and, thus, could contaminate the organization's IP portfolio.

- **Motivating and Rewarding**

  Many corporate reward programs are not structured to recognize and reward innovators. Therefore, special reward and recognition programs for corporate innovator should be developed to not only retain innovative talent but also to attract new talent.

The advantages that a corporate innovator has over an entrepreneur in commercializing technologies are significant. Most of these advantages are utilized by employing the corporate assets and departmental staff when implementing innovation. The disadvantages of corporate entrepreneurial activities are primarily related tithe corporate culture and management style. The corporate culture, hierarchical structure, bureaucratic management style, punishment of risk taking, faulty reward systems, and turf wars have to be dealt with firmly by top management and by the corporate innovators before accepting the responsibility of pursuing entrepreneurial activities. The entrepreneurial spirit does not become a major part of the corporate culture overnight. It takes a plan, a process, and leadership from the top to develop this in any corporate environment.

SUMMARY

- Life science innovators may dream of striking out on their own, but this is not always necessary.
- There are significant advantages in working with your employer to commercialize you technology.
- If an innovator understands their corporate culture and politics, he or she should carefully weigh the pros and cons and then make an educated decision about whether or not innovating within an organization is the right thing for them to do.

Chapter 12:

Commercialization Strategies

A key challenge for all researchers is determining how to translate their promising technologies into new products and services. Thinking about this early on and then developing a sound commercialization strategy involves being able to transform your new discovery into an enticing value proposition for all stakeholders. Keep in mind that the initial commercialization strategy may change over time. The commercialization strategy that is initially selected could change because of:

- more data collected about the technology,
- increased knowledge about the market,
- changes in the competitive landscape,
- inability to obtain IP rights,
- failure to receive FDA approval,
- failure to obtain reimbursement, and
- changes in personal situations.

Commercializing bioscience technologies is another key challenge and truly takes a community which further complicates getting the technology to market and developing a strong commercialization strategy. Another issue is that each type of medical product/service has its own timeline and different regulatory requirements. The evolution of a medical concept into a product requires an appreciation of challenges at each development phase described earlier. If the technology is licensed from an academic institution, the design of a medical product will be somewhat dependent the involvement of the inventors sharing their technical knowledge. Securing the necessary funding will be critical getting the product ready for the marketplace. At the later stages of development, support for selling to the customer and expanding the market might require industry partners.

Commercialization strategies for medical products involve substantial risks and present distinct benefits. Therefore, your commercialization strategy must be carefully thought-out, planned, and executed. An iterative process must be used as more information is collected as you assess the technology for commercial potential, develop a strong business model, and strategize commercialization strategies.

LIFE SCIENCE COMMERCIALIZATION STRATEGIES

There are several primary ways bioscience entrepreneurs can commercialize their inventions. Effective commercialization strategies typically present a trade off between:

- licensing your invention,
- working with your current employer and their commercialization managers,
- establishing a new venture around a technology and competing against established firms,
- selling your IP rights,
- forming a strategic partnerships, or
- creating a joint venture.

The most common commercialization strategy is to license a technology to another entity willing to further develop it, market it, and sell it. Each of these strategies is elaborated on below.

LICENSING A BIOSCIENCE INVENTION

A license is a formal agreement granting permission by an owner (the licensor) to a company (the licensee) to use the owner's proprietary information, invention, or material. A license can be granted for patents, copyrights on literary or artistic works, trademarks, and trade secrets. A licensing agreement gives the licensee a limited right to reproduce, sell, or distribute what is licensed. Intellectual property licenses usually require that the licensee pay a fee to the licensor in exchange for use of the property. When licensing intellectual property, neither ownership nor title of the IP changes.

As an example, Dr. Jones invents a medical device. After doing a risk analysis, he decides that the market for the device is not big enough to justify creating a company to manufacture and sell the instrument. He approaches an existing medical device company that has expressed an interest in co-developing the idea. Dr. Jones agrees to license the technology to the medical device company in exchange for payments when set milestones are reached and a share of the royalties on all sales.

A licensing strategy is often selected by an organization who owns the IP rights from their employee's invention. However if the inventor owns the IP rights outright, he/she has the right to license the use of the intellectual property to another entity to manufacture, market, distribute, or use it. Royalties can be negotiated on all sales of the product that is produced using the IP, and an up-front fee in addition to the royalty fee may also be negotiated. Another option is to license only the marketing and the manufacturing of the invention.

When considering licensing a bioscience technology, follow these helpful guidelines:

- Decide exactly what you intend to license (a product/ service, the design, a process, and so on).
- Define the benefits the IP provides to a potential licensee. Why would a potential licensee want to license the IP?
- Research potential licensees to ensure they have

the resources to properly use your IP and that their reputation in the industry is sound. Do not license your IP to a company to a weak or failing company with a poor reputation.

- Decide whether you will license the IP as it is now or allow the licensee to make modifications.
- Determine for how much you want to sell a license for.
- Consider including a confidentiality clause as part of the licensing agreement defining who may know the secrets of your IP and to whom they may disclose the information.
- Determine the length of the licensing agreement and specify whether the licensee can sublicense your IP or assign the rights away.

Because licensing is a legal transaction and licensing agreements are legal contracts, we strongly recommend that you consult an experienced attorney who specializes in licensing, not only to protect yourself against unethical licensees or licensors, but to make sure you negotiate the best deal.

Licensing a life science product does not mean instant success. The same due diligence should be performed on potential licensees as you would starting a new business. Detailed planning, good selling, and effective marketing are all needed for success. Additionally, the transfer of knowledge from the licensor to the licensee can create many challenges. For example, an inventor and his team possess great knowledge and know-how about the technology, the potential market, and manufacturing issues. When the technology is licensed, the knowledge and know-how needs to be transferred to the licensee so they can effectively sell and/or manufacture the product. Often, as part of the licensing agreement, the inventor and the team liaise with the licensee to ensure this knowledge transfer occurs. Without successful knowledge transfer, the licensee may not be successful.

Trade shows and association meetings are places where you can meet potential licensees, distributors, and/or sales representatives to learn about potential licensing opportunities. Other resources for licensing technology are colleges, universities, corporations, nonprofit research institutes, and information services such as the "American Bulletin of International Technology," "Selected Business Ventures" (published by General Electric), "Technology Mart," "Patent Licensing Gazette," and the National Technical Information Service.

Another good resource is the International Licensing Industry Merchandiser's Association (www.licensing.org). Another key resource is *The Licensing Resource Directory*, published annually by Expocon Management Associates, which lists owners and product/service categories as well as tips, support organizations, consultants, designers, and other professionals specializing in licensing. *The Licensing Letter Resource Sourcebook* from EPM Communications lists the top 4,000 executives involved in the licensing industry.

STARTING A NEW VENTURE

Creating a new venture, whether it is a spinout from a corporation or university or independently to commercialize a product, is a typical entrepreneurial pathway. However, it involves more risk, time, and effort compared to licensing a technology or joining with a strategic partner. If bioscience entrepreneurs are considering creating a new venture to commercialize their inventions, success and profitability depend on several factors including:

- protecting their intellectual property;
- developing strong intellectual protection barriers;
- assembling a strong management team to launch and grow the venture;
- creating a unique customer value proposition;
- focusing scarce resources on gaining market presence;
- securing funding to complete a customer prototype;
- crafting a competitive pricing model;
- generating a sustainable competitive advantage;
- obtaining funding for clinical trials; and
- securing industry contacts.

Other problems and challenges could arise if the bioscience entrepreneur discovers that the technology solution is unworkable, not needed, could quickly become obsolescent, requires a large capital investment, or takes a long time to achieve profitability. This is why the licensing option is often selected as a commercialization strategy.

Achieving success with launching a new venture also depends on the competitive strategies of existing companies, such as aggressive pricing wars and their ability to quickly imitate the functionality of your technology. The management team will also be challenged to build a rich stream of economic returns for founders, investors, key stakeholders, and staff.

STRATEGIC PARTNERSHIPS

Many bioscience entrepreneurs decide they do not want to assume all the business functions, financial commitments, and risk required to commercialize inventions. Instead they chose to form a strategic alliance with another entity, often in their supply chain. A strategic partnership is a formal alliance between two entities that each possess a skill set that the other wants to tap. Strategic partnerships are not legal partnerships. For example, a strategic partner could help further develop the technology for the market, manufacture it, market it, and the like.

Another example of a strategic partnership is when a healthcare professional invents a bioscience product and starts a new venture outside of his or her job. Then the inventor and his/her employer forms a strategic partnership where the inventor performs all the business functions of getting the product to market but is able to utilize the employer's equipment or as a site for clinical trials. Regardless of how the strategic partnership is formed, an experienced attorney should draw up a formal agreement that includes exit provisions, pricing formulas, IP ownership, and how to handle disagreements.

JOINT VENTURES

Another option is creating a joint venture where two or more parties, usually businesses, form a new venture to share markets, intellectual property, assets, knowledge, and, of course, profits. The forming parties agree to build a new entity together by each contributing equity and then sharing revenues, expenditures, and management of the enterprise. The joint venture could be for one distinct project or a continuing business relationship. The joint venture, most likely, operates independently from either of the joining entities.

Joint ventures can be created by large corporations or by small businesses that believe partnering will help them successfully fight bigger competitors. In some cases, a large company can decide to form a joint venture with a smaller venture in order to quickly acquire critical intellectual property, technology, or resources otherwise hard to obtain, even with plenty of cash at their disposal. Likewise, companies with identical products and services could join forces to penetrate markets they would not or could not consider without investing tremendous resources. Lastly, competitors could also form joint ventures to expand and better penetrated the marketplace.

There is no set ownership structure in joint ventures. Ownership interests range from a minority holding to a 50-50 split to a majority holding. Sometimes if a larger organization is teaming with a small business, the larger organization provides the funds to finance the joint venture while the small business provides the time, effort, and drive.

Overall, a joint venture can also speed up the time to market, break down barriers to entry, and increase market share. They could also be used to improve agility of the individual partners, provide diversification opportunities, and/or support transfer technology of one entity to the other. Regardless, there are three primarily considerations when forming a joint venture which are:

1. choosing the right partner,
2. deciding the ownership structure, and
3. determining who will manage the joint venture.

A joint venture agreement should be developed and provide for exit strategies specifying the conditions under which the parties can exit. It should also include pricing formulas to be used in the event of buyout as well as the process for handling unresolved disagreements.

ALTERNATIVE ENTREPRENEURIAL PATHWAYS

CONSULTING TO OR WORKING FOR INDUSTRY OR OTHER SERVICE PROVIDERS

Healthcare professionals, scientists, and engineers frequently work as consultants to industry and early stage companies. They perform various roles helping with product development, strategic planning, sales, and marketing and finance. By serving as advisors, members of scientific or business advisory boards, the management team, or as a member of the Board of Directors, consultants contribute content and domain expertise, credibility, and connectedness.

RESEARCH AND DEVELOPMENT COLLABORATIONS

A common mechanism for healthcare employees or academics to work with industry is partnering in research projects or cooperative research and development agreements (CRADA). Typical examples are performing experiments or tests at a university lab, participating in clinical trials, or serving on data quality and integrity committees.

KNOWLEDGE EXCHANGE AND KNOWLEDGE TRANSFER PROGRAMS

Knowledge transfer programs (KTP) are designed to cross fertilize academic environments with industry participants and vice versa. The goal is to give market direction to basic research projects, foster trust between academics and industry, stimulate industry investment in outsourced basic research, and provide work experience to trainees interested in working in the life science industry.

OTHER OPTIONS

Other options include investing as an angel investor or working for a private equity firm or investment bank for a venture capital fund. Participating in developing policy and helping to develop clusters working with local, regional, national or international biocluster developers are also good options as well.

12 CRITICAL COMMERCIALIZATION STRATEGY QUESTIONS

Your personal and business intentions will strongly influence which commercialization strategy you chose as well as the market, regulatory, reimbursement, economic and competitive conditions, not to mention the funding requirements to develop and commercialize a bioscience technology. Critically consider the twelve critical commercialization strategy questions in Table 1 to help decide the best commercialization strategy for your bioscience innovation.

The last two questions can be especially problematic. Every new technology entering the market needs some breathing room from competition to allow it time to develop and gain acceptance. Typically, technologies rely heavily on intellectual property protection to provide room for incubation and growth. A bioentrepreneur's commercialization strategy depends not only on the strength of intellectual property protection but also on the threat of competitors seizing or exploiting the invention's underlying technology.

Table 1: 12 critical commercialization strategy questions

1.	What are my vision and goals for the technology?
2.	Does my personal situation affect whether I license it or start a new firm?
3.	Do I have the skill set (or willingness to develop those skills) necessary to launch a new venture?
4.	Is there a short window of opportunity to get this technology to market?
5.	What is the economic environment?
6.	What is the industry's competitive environment?
7.	What will be the cost to enter the target market?
8.	What are regulatory and reimbursement requirements to consider?
9.	What are the industry's current channels of distribution?
10.	What are the revenue projections for the technology?
11.	What is the strength of the intellectual property protection?
12.	Could my invention preclude existing firms with prior knowledge about the technology and the industry from developing a competing technology?

SELECTING A COMMERCIALIZATION STRATEGY

If your employer has ownership IP rights to your invention, you will not develop an independent commercialization strategy but instead will work in concert with your technology commercialization office. Bioentrepreneurs who work with their organization's commercialization professionals will learn what it takes to get a new bioscience technology to the marketplace. In the future, they can better decide whether they are cut out to start a new venture on their own.

Another option for academic and corporate entrepreneurs is licensing the technology from their organization by either producing or selling the product themselves, licensing parts or all of the manufacturing, sales and distribution, and so on. They could also collaborate with a strategic partner to fund some of the development for the marketplace.

On the other hand, if you hold the IP ownership rights to your technology, it comes down to this key question — should you license the technology, form a strategic partnership or joint venture with another party, or should you sell the IP to someone else. Use Table 2 to determine which commercialization strategy is best for you and your technology.

Regardless of the commercialization strategy selected, bioentrepreneurs still need commercialization training, mentoring, and funding to successfully take inventions to the marketplace. Your technology's potential success lies with you even if you have access

Table 2: Technology commercialization strategies for ip owners

Commercialization Strategy	Risk	Revenue Potential	Management	Control
Sell IP	None	Very low – High	None	None
License	Low – Medium	Low – High	Low	Low – Medium
Strategic Partnership	Medium	Medium – High	High	Medium
Joint Venture	Medium – High	Medium – Very High	None – High	Low – High depending on joint venture agreement
Start-up	High	None – Very High	High	Low – High depending on equity investors

to your organization's technology commercialization office's valuable resources.

IN SUMMARY

- Develop a sound commercialization strategy from the outset.
- Commercialization strategies will be modified or changed during the evaluation and vetting process while pursuing IP protection as you learn more about the market, your industry and competitors.
- Traditional commercialization strategies include:
    - Licensing the technology,
    - Launching a new venture,
    - Selling the IP outright,
    - Forming a strategic partnership, or
    - Creating a joint venture.
- Select the best commercialization strategy for your invention that meets your personal goals and objectives.

COMMERCIALIZATION STRATEGY RESOURCES

Society of Physician Entrepreneurs (www.sopenet.org)

Gans, Joshua S. and Scott Stern. The Product Market and the Market for "ideas": Commercialization Strategies for Technology Entrepreneurs. 2002, Intellectual Property Research Institute of Australia.

Stim, Richard. *Profit from your Idea: How to Make Smart Licensing Deals*. NOLO, 2011.

Chapter 13:

Presenting your Business Ideas

The ability to communicate the value of your idea to others is critical whether it is to decision makers, customers, partners, employees, or potential investors. New life science innovations can look great on paper, but without clearly articulated and effective presentations, investors will most likely not be interested. Whether you have thirty seconds or thirty minutes to present, you must be "pitch-ready" to gain support, approval, and/or attract funding.

Putting together winning presentations is an art and gaining support for your opportunity is a critical doorway to success. Most innovators feel they can present themselves in a professional and convincing manner, but it does not happen easily. Most innovators, especially those with technical backgrounds, are not familiar with how to communicate their opportunities to decision makers who are more interested in the business aspect and usually do not have a technical background. However, you must be able to inspire others, get them excited about your opportunity, and solicit their support to move it forward. Developing presentations takes preparation, design time, and lots of practice.

The presenter must be convincing and display leadership ability and passion. Adding humor to the presentation can enhance it and further engages the audience. Remember that investors and decision makers bet on the team; the technology is secondary.

ARTICULATING YOUR IDEA

While there are unique challenges and characteristics of new bioenterprises, fundamentally the elements of success are the same as any other business. All too often, an inventor or scientist discovers or invents something, and then they look for a problem that their technology could solve. A more successful strategy is to find a problem, design a solution that is a significant improvement over what is currently being offered, and then execute a plan to develop and commercialize it.

Once you identify a large, growing, and potentially profitable problem, you must define a value proposition that describes what value your solution offers potential customers. Value is the difference between the tangible and intangible benefits of buying a product or service and the tangible or intangible costs. Your value proposition is the promise you make to your customers to provide them value if they buy your product.

Decision makers and/or potential investors also want to be convinced that you have developed a business model that will generate profits and sustain an advantage over competitors. The value proposition and business model will need to be better and stronger than the competition. Otherwise, there is no reason for customers to switch from what they are buying now or using the status quo to your products, and decision makers and/or investors will not get excited about your opportunity.

THE QUICK ELEVATOR PITCH

The elevator pitch is one method of communicating your idea to potential investors or those who are in a position to support it and help move it forward. The elevator pitch is a brief overview of your opportunity and quickly highlights its most important points.

The elevator pitch gets its name for being able to explain an opportunity and its value in the time span of a short elevator ride. An elevator pitch can be any type of short presentation that presents an idea, promotes a new venture, or markets to an individual. It can be as short as 30 seconds or as long as several minutes. Today, everyone should know how to develop and effectively present an elevator pitch since we always need to explain our new idea and obtain support from key decision makers to move forward.

No matter how long you have to give your elevator pitch, it is important to precisely explain your idea, what problems you are solving, your market opportunity, and your value proposition. It is your chance to shine and make a favorable first impression. The goal is to engage the audience and secure a follow-up meeting with interested parties or win approval to move forward with your idea.

Your elevator pitch should be non-technical, concise, and engaging. It should pique curiosity and be modified to represent the interests of your audience. Focus on the benefits of the opportunity which might include reduced costs, time-savings, efficiency, accuracy, increased safety, and so on rather than just giving a description of the idea.

Avoid trying to explain your opportunity in technical terms barely understood by a subject matter expert. Instead focus on what your audience is interested in. Your grandmother should be able to understand your opportunity and the point of your pitch.

The development of your feasibility, commercialization, or business plan will result in a clearer and more precise, compelling value proposition as you go through several levels of analysis. When you deliver an elevator pitch, you should clearly and succinctly answer the following questions:

1. What is your idea?
2. What is its stage of development?
3. What market(s) does the business address, and do you have any testimonials or customer feedback?
4. Why do you believe you have an advantage relative to market needs?
5. Who is the competition in the marketplace?
6. How will you make money?
7. Who is the team that will make the business succeed?
8. What is the long-term vision, *i.e.*, the exit strategy, and the projected return on investment for investors? (some examples are: "The business will distribute profits to

investors from cash flow by year X", or "The business will be acquired by another company for $Y")
9. What is the total funding required to execute the plan?
10. What amount of financing are you seeking initially, and, if looking for investors, what are the terms of investment?[1]

One model to use is the following outline. Just fill in the blanks:[2]

- For (*target customers-beachhead segment only*)
- Who are dissatisfied with (*the current market alternative*)
- Our product is a (*new product category*)
- That provides (*key problem solving capability*)
- Unlike (*the product alternative*)
- We have assembled (*key whole product features and benefits*).

Another similar model is the N.A.B.C. model:[3]

- Define the NEEDS of the target segment
- Say what APPROACH you are taking to eliminate the market pain
- State clearly the BENEFITS /Cost of using your product
- Differentiate your product from the COMPETITION

THE PITCH

Perfecting the art of presenting your feasibility or business plan to investors and other stakeholders, pitching, is a core entrepreneurial skill. Pitching is a combination of selling, persuading and educating your audience. Your goal is to generate interest by telling your

1 http://www.startupnation.com/pages/contact/elevator_pitch.asp
2 Moore, G. Crossingthe Chasm: Marketing and selling high-tech products to mainstream customers. *Harper Business*, Revised edition. 1999
3 http://www.cxacademy.org/developing-better-value-propositions-using-the-nabc-framework.html

business story so that it will appeal to the pocketbooks, hearts, and minds of investors and ask for a call to action, *e.g.* contacting you to learn more, asking for a product demonstration, or investing money in your company.

While the elevator pitch is just a couple of sentences, the feasibility or business plan pitch, is delivered in 15 to 20 minutes and is usually supplemented with a visual presentation, like a PowerPoint slide deck or other presentation graphics.

Pitches typically have three parts:

- the hook,
- the story, and
- a call to action.

The hook sets the stage. Your first words should grab the audience's attention, make them want to learn more, and succinctly describe what your idea does and why the audience should care. The language should be as free of "technospeak" or "business babble" as possible and explain your business ideas. For example:

- "Oncolight is a medical device company creating a tool that will help doctors optically detect cancer."
- "Omnibiopharma is a drug company developing a new treatment for anthrax."
- "Medvoy is a globally integrated doctor to doctor referral platform connecting physicians around the world."

PITCHING BEST PRACTICES

Becoming "pitch ready" involves the following three phase process.

1. **Preparatory Phase**

   Begin by researching your audience to get a better understanding of them: what is most likely to interest them, what they are looking for, and who will be present. Each audience will have different sweet spots, investment philosophies, and goals. Tailor your presentation based on the audience so that you push the right buttons. Through your research determine:

- Who will be there?
- What do they want to hear?
- Why do they want to hear it?

For example, an angel investor in the audience might have a family member, relative, or friend who has a disease that you are targeting. Understand and try to match the stage of your company and financing needs with the audience. Friends, family, and angel or high net worth investors look for different things than private equity or venture capital investors. Each has a different risk tolerance and will require more or less clinical validation, Phase 1 or Phase 2 clinical trial data or some demonstration of proof of concept. In addition, investors in later stage companies typically want to raise a minimal amount of money to justify their efforts on your behalf and will require that you relinquish varying levels of control of your company in exchange.

The biggest public speaking mistake people make is not practicing their presentations enough. Prior to the main event, be sure to rehearse your presentation several times in front of an honest, critical audience and make the necessary changes.

2. **Delivery Phase**

Arrive at the site about 15 minutes early to check any required technology, get a feel for the seating arrangement, and confirm how much time you will have for your presentation. Be sure to bring copies of anything you want distributed, having sufficient supplementary marketing materials should they be requested, and always have a back-up plan should you encounter technical difficulties.

When delivering the pitch, use the 10/20/30 rule[4]: ten slides, twenty minutes, thirty point fonts. The purpose of the pitch is to stimulate interest, not close the deal. As such, the pitch should be succinct and memorable and make listeners want to act or know more. Below is a template the authors have developed to use as a sample pitch slide deck.

Slide Contents	Discussion
Name of Opportunity **Names of Team Members** **Date**	• If appropriate, thank the person introducing you • Briefly introduce yourself and your team members • Introduce your opportunity
The Problem • Tell a story to hook the audience • For (target customers) • Who are dissatisfied with (current market alternative)	• Paint of picture of the burning customer problem • Tell a story – "Have you ever…?" • Explain the compelling customer need
The Solution • Our product/service is a (new product category) • That provides (key problem solving capability) • Unlike (product alternatives) • For this market opportunity • We have (key benefits of the product/service)	• Explain your solution to the burning problem • Focus on the benefits your opportunity provides customers • Why is your "pain killer" a better solution? Why is it compelling?
Risk Factors • What are the risks involved with this opportunity? • What are your mitigation strategies for each risk?	• There are always risks with any business/investment opportunity • Provide assurance you have identified risk factors and are prepared to address them
Business Model • How will you make money with this opportunity? • Highlight critical business alliances	• Discuss how you will make money doing what you say you will do • Explain the business logic of your opportunity • Describe key alliances • Give examples of other similar business models successfully used in the same industry

4 Kawasaki, G. The Art of the Start, *Portfolio*, 2004

Product/Service Overview • Purpose and unique features (why is it desirable) • Stage of development (working model, prototype?) • Status of intellectual property • Barriers to entry • Fields of use for the opportunity (initial users? Future users?)	• Show pictures of the product/service • Briefly describe product/service features but emphasize customer benefits • Explain IP protection in place or status of filed IP rights • What is your "secret sauce"? • Describe the barriers to entry • Show prototype is appropriate
Market Opportunity • Market size and potential growth • Industry trends • Customer profile • Target market niches	• Describe the market size and acceptance of your solution • Provide credible, well-founded market statistics • Validate all market growth projections with reliable data • Cite all sources
Sales • Sales strategy • Sales projections • Sales cycle • Pricing strategy • Current sales to date	• Identify if you have a super salesperson on board • Explain your sales strategy and why it will work • Mention any proven sales strategies • Summarize sales projections and explain assumptions • Describe your sales cycle • Introduce your pricing strategy • Emphasize current customers
Competition • Competition profile • Sustainable competitive advantages • Points of differentiation • What's your competitive edge?	• Describe what you have learned from competitors highlighting their strengths and weaknesses • Address threat of substitutes • Focus on your competitive advantages • Discuss how your opportunity is differentiated from competitors • Explain how you can sustain your competitive advantage
Management Team • Highlight current team's previous experiences and expertise • Explain what roles each member plays • List key talent that needs to be recruited and include a timeline	• Describe relevant background, experience, and expertise of team members • Explain strong track records • Show how past experience qualifies the team to run this venture • Do not ignore identified weaknesses in the management team or positions that need to be filled
Financials • Gross margin • Breakeven point • When will the company be profitable? • How much have founders invested to date? • Development needed to commercialize?	• Explain your funding strategy • Describe key financial ratios • Identify the amount and timing of funds needed for future growth • Mention key assumptions underlying financial projections

Summary • Value of opportunity • Explain why audience should invest or give you approval • Action steps • Summarize growth plans **Questions?**	• Develop a strong, memorable close • If audience is investors, tell them how much money that will make • If audience is decision makers, explain how the opportunity will add $X to their bottom line, profit, or help promote their mission • Show that you have an exciting and strong future ahead • Thank audience for their attention • Ask for questions

3. **Post Presentation Analysis Phase**

   Afterwards, evaluate the presentation by obtaining feedback, taking notes, answering questions, and deciding what worked, what did not, and how you could improve or modify it. Think of how you had planned the pitch to go, how it actually went, and how you would do it differently next time.

   Be sure to follow up on any promises you made to provide additional information or resources and thank your audience for their time and consideration.

TIME CONSTRAINTS

You will always have more to say than time to present your opportunity. It is critical to know how much time has been allotted for your presentation. You need to plan ahead so you can deliver it in the time allotted without being rushed. If you go over your scheduled time, you may risk being stopped before you have completed your presentation. Focus on the most important points during your time which will give your opportunity the best chance of being supported. If your presentation is too long, shorten the technical portions first, and avoid giving too much information about every aspect of the opportunity. Typically, innovators spend too much time trying to "wow" the audience about the technical aspects of their product or service instead of describing the value proposition and laying out the business model.

QUESTION & ANSWER SESSIONS

Allow some of your allotted time at the end of your presentation for the question and answer session. An important part of your presentation is being able to succinctly and quickly respond to the audience's inquiries. This time also allows you to build rapport, excitement, and enthusiasm about your opportunity. Anticipate questions that will be asked and prepare responses ahead of time.

When answering questions, repeat the question asked before responding. This will give you time to ensure you understand the question and formulate and articulate your response. Be direct and to the point when giving your response. If you cannot answer a question, say so, and then explain that you will obtain the requested information and will supply it later. Always follow through on your audience's requests. Remember to thank the audience members for their attention after all questions have been answered.

RED FLAGS AND PITFALLS TO AVOID

There are key pitfalls which can trip you up from either obtaining approval to move forward or attracting funding. For example, some innovators place unreasonably high valuations on their opportunity without valid statistics to back up their projections. If an opportunity has no revenue, no customers, barely a proof of concept, a market with many competitors, or a management team heavily weighted with technical capabilities, investors will most likely run from you. If you have already secured some funding and do not have much to show for it, most likely your opportunity will be treated with great skepticism.

Another red flag is an opportunity which lacks customers. This is problematic for a new early stage product/service to the marketplace since often there is not a working prototype. However, it might be possible to get pre-purchase orders or letters of endorsement from potential customers. Another sure way to ruin your chances of having a successful pitch is being unprepared to make your presentation and/or stumble during the question and answer period. Even the most appealing opportunity will not attract interest if your presentation is disorganized, poorly thought-out, or exceeds the allotted time before major points are covered.

Seasoned presenters avoid using the following phrases:

- "Well," "Umm," or "OK" to begin your presentation;
- "No one else does what we do";
- "Key employees will join us at funding";
- "Bigger competitors will be slow to react";
- "All we have to do is get one percent of the market";
- "This is a billion-dollar market";
- "This is the best deal you will ever see"; and
- "Several VCs are very interested already".

KEYS TO SUCCESS

Below is a short list of the most important keys to success when communicating to key decision makers, potential investors, and/or customers.

- Be yourself;
- Smile;
- Be professional and convincing;
- Always tell a story;
- Focus your presentation on the impression you want burned into the audience's minds;
- Move around the room — don't be tied to one spot;
- Don't turn your back to the audience for any reason;
- Maintain eye contact with the entire audience;
- Speak slower than normal — presentation nerves causes us to talk faster;
- Communicate your excitement for the opportunity in your voice; and
- Speak loud enough so the person in the back of the room can easily hear you or use a microphone.

SUMMARY

- Pitches articulate your idea's value proposition and describe what you want from the audience.
- Create a short, succinct elevator pitch describing your business.
- Practice your investor pitch, get feedback from honest critics, and constantly improve it.
- Sharpen your public speaking and presentation skills.

Chapter 14:

Financing Your Ideas

The challenge for bioscience entrepreneurs is to translate their scientific discoveries into valuable intellectual property and then monetize it through a license, acquisition by another company, or an initial public offering (IPO). The value of bioscience ventures and their IP steadily increases as they progress through proof of concept, clinical trials, and FDA approval to finally the creation of a marketing product.

Understanding financials is vitally important and the challenge for moist bioscience entrepreneurs is they do not have strong financial backgrounds and financing their inventions or new company is foreign to them. This chapter is divided into two sections. First is introducing funding sources for life science inventions and the second section briefly reviews important finance statements need to acquire funding to further develop the technology for the market.

FUNDING SOURCES FOR LIFE SCIENCE INVENTIONS

Where do bioscience entrepreneurs find money to further develop and take their technologies to the marketplace? Usually, several stages of investment are required over the life of a company and/or getting a technology ready for market. Different funding sources are used for research funding, start-up capital, working capital, and growth capital. During early R&D stages, the value of the invention is relatively low making it difficult to attract equity investors. They generally prefer later-stage venture with more clinically-advanced bioscience products where the risk is lower. This is the reason why many bioscience entrepreneurs work with their employers and utilize their internal funding sources to get their invention market ready.

While developing a financial strategy, the bioentrepreneur must consider the types of financing that are available and he/she is comfortable using. There is a variety of debt, equity, and non-traditional or bootstrapping sources that can be tapped depending on:

- the stage of development of the invention,
- the market need,
- the attractiveness of the invention, and
- a convincing business plan.

DEBT FINANCING

Debt financing consists of loans where a creditor, such as a bank, agrees to lend money to a borrower in exchange for repayment with accumulated interest at some future date. Debt financing is attractive because the borrower does not have to give away ownership in the venture. Some advantages of debt financing are the financing costs (*i.e.* interest) are relatively low, the loan interest is sometimes tax deductible, and the payment is a fixed monthly expense.

Obtaining a debt financing is usually faster than identifying equity investors, finding those who might be interested, trying to convince them to invest, and then negotiating a deal. However, debt financing is usually used by existing ventures with a strong financial track record and corporate assets or by start-up ventures with modest cash needs. A potential start-up venture will most likely need to have collateral to pledge in order to secure a business loan.

Typical sources of debt financing are:

1. Banks and credit unions are good for ventures with an established operating history based on credit records, ongoing operations, and/or physical assets to use as collateral.
2. The Small Business Administration provides

government-guaranteed loans made by commercial lenders. The criteria to secure a business loan are less stringent than a traditional bank loan, but the terms are usually more expensive.

3. Credit cards, second mortgages, and other consumer debt are often used for short-term, small cash requirements, wherein the venture can quickly pay off the loan.
4. Friends and family can be tapped when the venture needs small amounts of cash and has the ability to make loan payments on a timely basis.

EQUITY FINANCING

Equity financing involves selling a partial interest in the venture to investors who usually receive stock in exchange for their money. In contrast to debt financing, equity financing does not involve a direct obligation to pay back the investment. Instead, equity investors become part owners and partners in the business and, thus, are able to exercise some degree of control over how it is run.

Most equity investors do not expect an immediate return on their investment during the early stages but bet on the venture being profitable in three to seven years. Equity investors can be passive or active. Passive investors provide funding but play little or no part in running the venture. Active investors may be employed by the venture and are heavily involved in its operations.

Equity financing is not inexpensive, as investors are entitled to a share of profits indefinitely. Conversely, some life science entrepreneurs who often find difficultly securing a traditional loan are willing to share control of their business with a partner who is willing to accept the risk of getting the technology to the marketplace. Also, this could be their only option. In these situations, inventors accept that equity financing can be a mutually beneficial arrangement.

In any event of business failure, creditors are usually paid before owners. Therefore, equity investors accept more risk than debt financiers. They also expect to earn a higher return on their investment. Because the only way they can recover their investment is to sell their stock at a higher price in the future, they are generally committed to furthering the long-term success and profitability of the company.

Equity Investors will invest in certain stages of company development with the expectation, like a relay race, that the baton will be passed to the next level of investment as the company moves forward.

Typical sources of equity financing include:

1. Friends and family members can provide equity funding for smaller early-stage ventures with marginal cash needs.
2. Angel investors provide equity funding for businesses with moderate cash needs, excellent growth potential, and a clear exit strategy.
3. Venture capital (VC) firms should be considered for high-growth larger ventures that have large cash needs, extraordinary growth potential, and a clear exit strategy of being acquired or going public.
4. Corporate venture capital funds from large corporations, such as Amgen, Eli Lilly, and DuPont, actively invest in early stage ventures.
5. Small business investment companies (SBICs) are privately-owned and management investment firms funded by the Small Business Administration.
6. Merchant banks provide another source of money that includes a combination of debt and equity financing. They usually are willing to sell back equity to the venture although their interest rates are usually higher than traditional banks.

BOOTSTRAP FINANCING

Bootstrapping involves a variety of clever methods to find and use existing resources and capital to get the new venture started. Bootstrap financing is a creative way and the least expensive means to manage expenses and start with as little cash as possible. It is an effective way to finance a new opportunity and hold on to all equity. In addition, it maximizes efficient use of capital and minimizes the costs associated with acquiring it.

Bootstrapping is a popular form of internal funding and relies on the team's ability to access all of their company's resources and free up any available capital to launch the opportunity. Besides being one of the most inexpensive ways to raise capital for a new venture, bootstrap financing is attractive to outside lenders if the time arises when it is necessary to acquire equity investment.

Typical types of bootstrap financing include:

1. Foundation, grant, and/or government funding does not require the venture to pay back the capital. It is funding typically used to develop proof of a concept or a prototype.
2. Customers can provide financing by using a letter of credit from them to purchase materials without using any of the venture's money.
3. Trade credit involves finding a vendor or supplier who is willing to extend trade credit and allow you to order goods on net 30-, 60-, or 90-day terms.
4. Bartering is becoming much more popular recently due to the global economic downturn. Bartering involves trading services or good between two parties rather than a good or service in exchange for cash. Typically bartering has been used to obtain equipment, inventory, vehicles, real estate, and professional services.
5. Leasing involves leasing equipment rather than purchasing it outright when cash flow allows regular lease payments to be made.
6. Purchasing used equipment rather than new equipment and materials can reduce the cost of buying required equipment.
7. Real estate borrowing generates cash by using internal assets if your venture is able to lease, refinance, and/or borrow against your equity.

NONTRADITIONAL FINANCING

There are many different types of alternative money sources that life science entrepreneurs should become aware of and learn how to use. Capital can come from your suppliers, customers, vendors, or anyone else involved with your business.

Typical sources of nontraditional financing are:

1. Professional advisors and business acquaintances have been known to invest in the enterprises they work with.
2. Suppliers sometimes can provide the cheapest money that can be borrowed and may go out of their way to help finance a new venture. Some manufacturers may allow "dating" which will extend the length of time for payment of inventory and asset purchases.
3. Small Business Innovation Research (SBIR) and Small Business Technology Transfer (STTR) programs provide government grants through various governmental agencies to small and high-tech businesses.
4. Cooperative Research and Development Agreements (CRADA) allow companies to partner with governmental agencies to help develop new technologies beneficial to both parties.
5. Bonds are a form of a note. It is not uncommon practice for bonds to be sold to investors in new ventures in combination with some kind of equity investment.
6. Customers could pay for merchandise or services at the time of order to assist the venture with financing while securing the needed product or service.
7. Equity stakes of a new venture can be purchased by another company to obtain the needed product or service and a reliable source of supply.
8. Joint ventures or strategic partnerships can be formed to share assets such as equipment thereby lowering capital requirements.
9. Vendors could furnish equipment and/or supplies in

return for an equity stake in the venture to secure initial and follow-on sales to the vendor.

10. Other creative arrangements of nontraditional financing are only limited by the creativity and resources of the management team.

FUNDING LIFE SCIENCE TECHNOLOGIES & VENTURES

Every new venture moves through stages of growth like human grow from infancy to early childhood to adolescence to adulthood. There are primarily four stages of growth for life science ventures and different funding sources are usually accessed during various stages of growth as described below.

STAGE 1: CONCEPTION AND GESTATION

During this first stage, the inventor strives to demonstrate scientific proof of concept or develop a prototype that is still in preclinical development. The initial funding for a new venture can come from several possible sources: government such as economic development or scientific research grants, organizations, universities, family, friends, angel investors, and the inventor. The amount is usually modest, in the range of $1 million to $5 million. This initial funding is rarely enough to do more than pay for studies to strengthen the scientific hypotheses. To move beyond this initial stage, the entrepreneur must develop a strong business plan and recruit a management team with industry experience and a successful track record.

STAGE 2: INFANCY

In this stage, the life science invention moves toward human clinical trials. This step requires the filing of an application with the FDA with strict and expensive regulatory requirements which must satisfied. These early studies can typically cost between $5 million and $10 million.

If initial investors are impressed with the venture's development progress, they may allocate more of their funds to the venture and maintain or increase their ownership interest. However, most investors are small and do not have all of the additional capital that the venture requires at this stage. Thus, the team must find new sources of financing to continue commercialization. Venture capital funds are typically sought during this stage along with large pharmaceutical companies' corporate investment funds. As the technology continues to make progress toward human clinical trials, opportunities may arise for the venture to license or to sell its IP to a company that will continue and complete the development process.

STAGE 3: ADOLESCENCE

Another watershed moment in the Life Science Innovation Roadmap™ is the FDA application for permission to begin clinical studies on human subjects. Getting to this point demonstrates great therapeutic potential which could increase the perceived value of the venture to prospective investors. This is matched by an enormous jump in capital required to cover the exorbitant cost of human studies.

The larger amount of required capital usually exceeds the financial resources of early-stage investors, including many VC firms. Yet, other funding sources must be contacted such as established pharmaceutical companies or another type of investment firm called a crossover fund. Large drug companies frequently establish special venture funds to invest in promising new technologies that complement their own R&D efforts.

The investors in this stage expect to earn back their investment in a few months to a couple years. The money is targeted toward the development of clinical data that can persuasively demonstrate the therapeutic value of the invention. The strength of the clinical data during this stage is a critical indicator to investors of the return that they can expect to earn on their investment. The amounts of additional capital required at this point typically ranges from roughly $10 million to $30 million, depending on the condition or disease the invention targets. In some cases, the necessary clinical trials are larger and more expensive, and more of them are needed to prove the concept.

Because of the larger sum of capital required and the looming risk of complete failure in the development process, willing investors are harder to find. The company's founders and owners may also be reluctant to surrender even more control of their business. As a result, some life science ventures try to negotiate with prospective investors to craft innovative financing packages. For example, entities like contract research organizations may offer modest amounts of financing supplemented by non-monetary investments like facilities, equipment, and development expertise. In place of equity, some investors may be willing to accept a legal right to future royalties the company earns on the invention.

STAGE 4: MATURITY AND INVESTOR EXIT

As the development process progresses toward having a marketable product, founders begin focusing on how to translate their imagination, hard work, and risk-taking into dollars. Simultaneously, the investors are planning their exit from the venture after earning substantial returns for their original investments.

At this point, there are two fundamental exit strategies:

1. Sell all or most of the equity usually through an IPO, or
2. Sell the entire business to another larger entity through a merger or acquisition.

The market for life science IPOs has slowed since its peak in the early 2000s. Some of the slack has been picked up by large pharmaceutical companies who are struggling to fill their R&D pipelines as their established "cash cow" products are coming off patent and facing the threat of competition from generic brands. Since 2005, over $60 billion has been invested in biotech company acquisitions by the pharmaceutical industry.[1] For example, Roche purchased the remaining 44% of Genentech for $44 billion in 2008.[2]The large pharmaceutical companies are also licensing the IP or acquiring new biotech company earlier in the development process to fill their R&D pipeline.

1 http://hc.wharton.upenn.edu/nicholson/pdf/merger_Sept03.pdf

2 Pollack, A. "Roche agrees to buy Genentech for $46.8 billion" *The New York Times*. March 12, 2009.

WHAT VCS LOOK FOR IN EARLY STAGE COMPANIES

As mentioned earlier, funding life science companies is more like a relay race than a marathon. During each stage of development, from the idea generation phase, to the feasibility and testing phase to the commercialization phase, entrepreneurs typically look to different kinds of funding sources and investors.

Large amounts of money are needed to do clinical trials, prototype design, fabrication, and testing. At this stage, venture capitalists can provide the dollars to finance the venture. Founders of early stage companies considering obtaining venture capital funding for development and growth, often find themselves wondering what venture capitalists look for when assessing an investment opportunity. The answer is neither simple nor is it exact. Nonetheless, there are a few factors that every venture capital fund will assess when looking to invest in an early stage company.

First, it is important to review the basics of how venture capital works. Venture capitalists seek investment opportunities that fit the fund's investment criteria. Over the last decade or so, venture funds have become very large and with limited personnel to tend to portfolio companies, the funds invest in only a handful of companies. Hence, the investment size for each investment is driven by the size of the fund. For example, a $250 million venture fund may only want to invest in 20 companies with an initial investment of $5 million per company and holding back another $7.5 million in reserve for further investment in each company.

In order to "realize" their return, the fund must ultimately liquidate portfolio holdings. Most venture funds seek to fully liquidate within five to seven years. To liquidate, the portfolio companies are urged to sell their company to a strategic buyer or take the company public. Many VCs think of returns in terms of an initial investment multiple. For example, they may expect to get five times their initial investment from a successful exit.

VCs take on significant risk by investing in early stage companies and will perform extensive due diligence beforehand. It would take volumes to discuss in detail what VCs look for, but generally, they will assess at least seven critical factors very closely. The better positioned a company relative to these factors, the higher the probability of raising venture capital. Figure 4 below outlines at these critical factors.

UNIQUE PRODUCT OR SERVICE

VCs will assess how the product or service is differentiated from competing products or services. Does it solve a real problem or fulfill an unmet need? How is it protected from competitors? VCs are very sensitive to IP protection rights, the more the better.

LARGE AND GROWING TARGET MARKET

Does the product or service target a large and growing market? A VC might assume that, at best, a company will capture ten percent market share.

SCALABLE REVENUE MODEL

Scalability refers to the ability to increase revenues at reduced costs to generate that revenue. Also, is it a recurring revenue model? An

Figure 4: 7 Critical factors for venture capitalists

example of a scalable and recurring revenue model is a telecommunication service provider. Once the infrastructure is in place, the business easily can add new customers onto the platform and customers pay recurring monthly fees for service.

SOLID BUSINESS PLAN

This is much more than just having a well written business plan. It is about the content of that plan. Does the company have a thorough, well-crafted strategy? What is the vision of the founders? How will management execute the plan? Is the plan feasible?

SIGNIFICANT GROSS MARGINS

High gross margins leave residual income to not only cover fixed costs but also provide a buffer for errors in forecasting and revenue downturns. Also important is how defendable are the margins. Is the business in an industry with low barriers to entry which invites competitors and restrict margins. Commodity products and services, such as office supplies, usually do not have high gross margins due to many strong competitors, pricing wars, and little control over the cost of goods. Software, on the other hand, tends to have high gross margins.

CLEAR EXIT STRATEGY

As indicated previously, VC funds need to liquidate portfolio holdings in order to get their money back plus any returns on their own investments. Private equity investments have a very limited secondary market unlike the public stock markets. VCs rely on portfolio companies being acquired or going public through an initial public offering to provide a liquidity event.

MANAGEMENT

The management team is the glue holding all these pieces together and making them work. A strong, seasoned management team is often the most critical element to raising venture capital. VCs will pay particular attention to management's experience, track record of success (not necessarily entrepreneurial success), education, ability to work as a team, leadership skills, and so on.

There are two additional concepts that are critical to all of the factors: adaptability and focus.

ADAPTABILITY

An emerging growth business must be adaptable. The management team must be able to quickly recognize challenges, embrace and change with new information, evolve the product, change the business plan and direction, adapt the revenue model, and adjust to ever changing regulatory and economic environments.

FOCUS

Aspiring entrepreneurs often want to set the world on fire. However, it is important to stay focused. For example, if your business is ready to roll out a product or service in the United States, focus on that goal. Do not try to roll out the product in multiple countries at the same time. Focus on the most critical applications of your product or service first to prove the model and then go after other applications.

PRO-FORMA FINANCIAL STATEMENTS

Most life science entrepreneurs enjoy working with numbers especially since much of their research involves collecting and interpreting numerical data. Approach finance and investment in the same way as you would approach numerical data in your field. Often, when analyzing financials, the numbers show trends which gives meaning behind the numbers.

Financial statements are tools as essential to business management as a wrench is to a machinist. Historical financial statements reflect various management decisions over a certain period of time. They measure business metrics and show how management decisions impact the financial health of the venture.

When first starting a new venture, historical financial information is not available to for financial forecasting required in both feasibility, commercialization, and business plans. Therefore, proforma financial statements which are future-focused and based on forecasts, predictions, and assumptions of how the venture or invention will make money, must be developed. These financial statements will help determine the financial requirements of the new venture. They will also help determine how much capital will be needed to operate the new venture which will impact what type of financing is pursued.

Pro-formas are never 100% accurate as they are only predictions. However, the goal is to forecast as accurately as possible based on current information. Pro-forma financial statements should be updated when new information is obtained or when more accurate assumptions can be made about the future. A business accountant is a helpful resource to use when developing pro-forma financial statements. Most pro-forma financial statements are developed for the first three to five years of the business.

The two primary business objectives of any new venture are profitability and solvency. Profitability is the ability to generate income and make money after all operating expenses have been paid. Solvency is the ability to pay debts as they become due. Three main financial statements are at core of all accounting and must be clearly understood by bioentrepreneurs. They are:

- Profit & Loss (P&L) Statement or Income Statement,
- Balance Sheet, and
- Statement of Cash Flows.

These three pro-forma financial statements are built upon each other beginning with the P&L statement. Each is briefly described below.

THE P&L STATEMENT

The P&L statement, or income statement, shows how much money the venture will make (or lose) for a given period of time. In other words, what is the venture's anticipated financial performance? P&L statements are usually generated for one year's time or shorter. Similar to a bank statement, the P&L reflects deposits (revenues), expenses, and how much money is left in the account at the end of the time period. Ideally, the revenues exceed expenses. However, it is not unusual for technology start-up venture to take several years before they generate any revenue.

THE BALANCE SHEET

The balance sheet shows the assets (what the business owns), claims to those assets by creditors (liabilities), and owners' equity. The balance sheet shows a snapshot of the business at a single moment in time. The balance sheet is based on the "Accounting Equation" which stipulates that the value of a business's assets must equal the value of the liabilities and the rights of the owners. Therefore, even though the assets are property of the business, both owners and outsiders (creditors) have rights to those assets. The accounting equation is:

This equation always holds true on the balance sheet.

CASH FLOW STATEMENT

The statement of cash flow is perhaps the most important financial statement for life science entrepreneurs. It is a critical management tool for decision making and controlling the venture's financial affairs. The cash flow statement carefully tracks the inflows and outflows of cash for the venture. The pro-forma cash flow statement may show a negative cash flow for the first couple of years which is not abnormal for technology ventures. The cash flow statement is broken into three categories:

- operating activities,
- investing activities, and
- financing activities.

The cash flow statement is similar to a checkbook register. You keep track of every check, debit card transaction, cash withdrawal, and deposit made so you know exactly how much money you have in your account. This is what the cash flow statement does for your venture. It records every cash entry into and cash payment out of the business. The cash flow statement helps with budgeting so there is enough cash on hand to pay bills.

PRO-FORMA STATEMENT ASSUMPTIONS

In the process of developing pro forma financial statements, hundreds of assumptions are made by hypothesizing what will happen in the future. It is critical to explain the rationale by including explanations as to how every assumption was calculated. Sound financial assumptions make the financials more credible to key stakeholders and/or potential investors.

For most start-ups, the most important assumption is the revenue projection. If this projection is incorrect, the remaining financial projections are invalid and have no value. Yet this projection is the most difficult to accurately predict. Most revenues do not start until several significant events have occurred such as product testing, building a customer prototype, FDA approval, manufacturing, and so on. Other important assumptions include cost of goods sold, payment and collection cycles, turnover, and headcount.

Avoid "hockey stick" financial projects which predict that the business will exponentially increase sales at an unrealistic rate. Due to high R&D costs, life science entrepreneurs often need injections of outside capital. Typically, this causes founders to inflate their financial projections so they are attractive to investors. Yet, when they do so, they cast doubt about the reality of these projections. Instead, assumptions should be based on both primary and secondary data reflecting realistic grow projections for the industry or similar industries.

Financial assumptions can either enhance or become a fatal flaw in a venture's chances of attracting outside funding. If the assumptions are realistic and backed with valid research, there is a better chance of attracting investors. More importantly, sound assumptions provide the best information about the potential success of a new life science venture.

Financial planning cannot be done in a vacuum and takes considerable time and effort to develop a sound fundraising plan and strategy for each stage of growth. Although you probably want to develop your invention because of its technical advantages or the contribution it could make, you must consider the cost-benefit or different alternatives available to you. If you are working with your current organization to commercialization your invention, it is equally important to develop financial projections of the amount of funding it will take to fully develop the technology to go to market.

SUMMARY

- Funding life science ventures challenges the most experienced team, particularly in difficult economic times.
- Learn about the different sources of financing for life science ventures.
- Develop a sound fundraising plan appropriate for your venture's stage of development.
- Learn to develop and use pro-forma financial statements to determine just how much money the venture will need and when it will be needed
- Develop a fundraising strategy early on to execute the business plan.

ADDITIONAL RESOURCES

Below are some helpful resources for you and your team as you begin to develop your funding strategy.

- Venture debt financing: Vencore Capital (www.vencorecapital.com)
- SBIR/STTR grants: www.sbir.gov
- Cooperative Research and Development Agreements (CRADA): http://www.fda.gov/ScienceResearch/CollaborativeOpportunities/CooperativeResearchandDevelopmentAgreementsCRADAs/default.htm
- Angels and angel networks: Keiretsu Forum (www.keiretsuforum.com)
- Strategic partners: major medical device companies (www.medtronic.com and www.gyrusacmi.com)

- Angel Networks www.angelnetwork.com
- Angel Capital Association www.angelcapitalassociation.org
- National Venture Capital Association www.nvca.org

CHAPTER 15:

Exit Strategies

When building a new life science venture, the management team should be planning at the very beginning how to increase the value of the venture and develop an exit strategy that will provide the return on investment the team and that its investors desire. Unfortunately, most management teams wait to see what the value the business has when they are ready to exit. Instead, the management team should develop an initial exit strategy early on event though it will periodically change as the venture builds and grows. In addition, attracting equity investors without a strong exit strategy for their investments is extremely difficult.

DEVELOPING EXIT STRATEGIES

Exit strategies are one of the most important components of starting a new venture and are critical to develop from the very start of a new venture. At some point, every business owner will leave their business — voluntarily or not.

An exit strategy is a plan for the founder(s) to capture the value created through the venture's business activities. It helps life science entrepreneurs define success for the new venture and focus on its eventual transition. It communicates what founds plan to do with the business in the end. It is important to decide whether the exit will be by acquisition; strategically sold to an industry player, supplier, competitor, or employees; or sold by an initial public offering. Your venture's exit plan helps determine its capital strategy as well as its entry, growth, positioning, and distribution strategies.

Life science entrepreneurs who do not plan for ownership transition are often faced with the inability to receive enough money to fund a comfortable retirement. This does not happen because they failed to create value in their ventures, but rather because they failed to plan how to leverage and build value for all of their hard work and effort.

Most equity investors require that the founder(s) have an exit strategy in place before investing. Investors are always looking for the venture's exit strategy to determine how and when they will make their money. Many investors will work with entrepreneurs to map out a proposed exit date at the time of investment. Likewise, they tend to shy away from ventures that do not have an exit strategy because it could be an indicator that the fund is more interested in building and running a "lifestyle" business rather than forming a potential high-growth venture.

Life science and physician entrepreneurs will have different exit strategies depending on their circumstances and goals. For example, a small to medium sized private practice or group practice might want to grow to satisfy the revenue needs of its owners. The exit would involve selling a share of the practice or the entire practice to a willing buyer or transfer ownership to a family member. On the other hand, those interested in building a scalable life science business or care delivery business, such as a company that provides intensivists to hospitals around the country or an emergency room staffing company, might want to go public via an IPO or be bought as part of a financial or strategic acquisition.

DEVELOPING AN EXIT STRATEGY

Developing an exit strategy forces you to consider your personal goals and those of the management team. The next step is to outline what success looks like for the founder(s), the management team, and the business. An exit strategy provides for contingencies to accommodate unforeseen events and circumstances. Developing an exit strategy is a process that requires careful thought and deliberation. Once it is initially developed, it will be continually refined and revises as business conditions and personal circumstances change.

Exit strategies typically include:

- Current valuation of the business,
- Key factors that drive business value,
- Methods to increase business value,
- Forecasted future value of the business,
- Ownership change options,
- Tax implications of ownership changes,
- Tax-saving methods,
- Specific exit goals,
- Exit timeline,
- Action plans to implement the exit plan, and
- Projected proceeds from an exit.

A successful exit strategy protects wealth, attracts valuable employees, and ensures a smooth transition. It allows the founder(s) to clearly communicate their goals and expectations to their team, stakeholders, and potential investors. No exit strategy, or a weak one, could mean financial ruin. Setting an exit goal and developing a strategy to achieve it is what separates successful entrepreneurs from the rest.

In addition, investors in your idea want to have a clear understanding of how and when you plan to exit before they invest. Their return on investment will depend on your exit assumptions and their validity. For example, you plan to develop a method for the nonsurgical treatment of a condition that now requires surgery. The business model and financial pro forma statements, projecting future revenue, outline your estimates of the growth of revenues, the value of the company as a multiple of those revenues, and, what investors could expect to receive should you sell the company in the future.

If the company needs to raise money to start the venture and more than one round of financing is required, the bioentrepreneur must decide how investors will get a return on their investment. Few privately owned companies want to pay dividends because doing so triggers double taxation. The company pays taxes on the profit of the company and then investors pay taxes when they receive dividends from the company. Therefore, attracting investors without a sound exit strategy for their investments is most difficult. Intense pressure will continue from investors for a well-defined exit strategy that will get them liquid within five to seven years of investing.

If the new venture is going to be housed under the corporate umbrella, then an exit strategy is not necessary. However, some companies encourage their employees to license technologies and start new ventures outside the parent corporation. The management team then negotiates a licensing agreement with the parent corporation and makes sure that the length of the license corresponds with the exit strategy.

THE EXIT PROCESS

There are 5 key steps to selling part or all of a business.

STEP 1: DEFINING OBJECTIVES

The first step involves defining your objectives, which, in most instances will involve how much money you want, who will retain control of the asset once it is sold, and your future involvement with the company, it any, after the sale.

STEP 2: FINDING THE RIGHT BUYER

There are many different kinds of buyers. Strategic buyers focus on capturing market share or consolidating the market. Financial buyers look for more short term opportunities to turn around a business and create higher profits for the owners.

STEP 3: PREPARING THE BUSINESS FOR SALE

Like getting your house ready for sale, you will need to get your business ready for sale to attract the highest offer. On the average it can take well over 3 years to get the venture for ale to outsiders. Overall, this involves streamlining costs, getting all financial documents and business records in order, and showing healthy increasing profits.

STEP 4: THE DEAL PROCESS

Once someone has expressed interest in your business, then the deal process begins, defining mutual interests, doing due diligence, verifying financials, and, hopefully, coming to a win-win agreement which can take months of time to finalize.

STEP 5: WHAT TO DO AFTER THE DEAL

Now that you have sold your company, how should you manage your earnings so you can minimize your tax burden, build future wealth and/or create an estate plan?

EXIT TIMING AND STRATEGY

Very few companies appear to exit following the classic venture backed model of investing early, developing a product, obtain regulatory approval, generate revenue growth, become profitable and then exit. Instead, exits can occur at various points in the pathway. Some have pointed out that the chance of a successful exit improves if there is competition for your assets, if you have done a good job or preparing and marketing your company for sale using investment banking help, if you align the venture syndicate, if you have an appropriate geographic market reach and are able to time the exit at a favorable point in the macroeconomic business cycle.[1]

STEPS IN SELLING ALL OR PART OF A BUSINESS

When buying or selling a business or part of a business often an investment banker is involved and the transaction typically is done in these five stages each described below: engagement, pre-marketing, marketing, negotiation, and closing phase described below.

Engagement

- An engagement letter is signed that defines the legal relationship between the investment banker or financial professional and the buyer or seller in an M & A

1 Buch, H.K., Gustafsson, A.C., Drvotal, V., Sundberg C.J. (2011) "Divining the path to a successful European exit" *Nature Biotechnology* 29:205-207

transaction.

Pre-marketing

- The investment banker representing the business works with the seller to develop a list of potential buyers.

Marketing

- The seller and banker create a teaser or a marketing piece that describes the business in general terms that is designed to stimulate interest from potential buyers.
- A confidentiality or non-disclosure agreement is signed when one company is planning to disclose confidential or proprietary information to another
- A confidential information memorandum (CIM) is use to describe in a much more detailed way than the teaser, information about the business and will include such things as an executive summary, an overview of the seller's industry and an overview of the products and services offered.

Negotiation

- An initial bid procedure is offered that indicates a deadline for prospective bidders and asks for information from buyers such as potential funding sources and the currency being used to buy.
- A data room is a secure confidential location where all parties in the deal gather to do due diligence relating to the transaction. Many firms use virtual data rooms instead of physical location.
- Management presentations are designed to provide additional information not included in the teaser or CIM. In essence, it's an opportunity for management to convince buyers why they should buy their business.
- An initial bid is not binding, but generally outlines the terms of an offer.

- The initial bids are considered by the seller, and responses are sent to those who have submitted offers that are not acceptable.
- Term sheets are created for the remaining bidders.
- The remaining bidders are given another opportunity to bid and a definitive draft agreement is included with the final bid procedures letter.
- The bidders submit their final offers with a firm purchase price.
- The seller selects the winner.

Closing Phase

- The final terms of the acquisition are be finalized in a written agreement.
- The selling company could request a fairness opinion from a third party to determine whether the proposed transaction is fair, particularly for those transactions that will require shareholder approval.
- Once the parties have agreed to the terms and conditions of the deal, they sign the closing documents to complete the transaction.

EXIT STRATEGY OPTIONS

The most common ways for a company to exit are going public, termed an initial public offering or IPO; selling the company; merging with another company; liquidating the business; reverse mergers; or bankruptcies which are briefly described below.

IPO

An initial public offering takes a company public by raising capital through federally registered and underwritten sales of the company's stock. It is a complex form of a private placement in which the founders and equity shareholders agree to sell a portion of their company (via unissued stocks and bonds) to the public by filing with the SEC and listing the stock on one of the stock exchanges. Once the company goes public, the management team becomes focused on maintaining the price of the company's stock by showing healthy profits and having the stockholders receive generous dividends on their stock.

Going public may be a terrific way for many technology ventures to raise substantial amounts of money, which they need for research and development, developing a working prototype, and taking prototypes to market over several years. Therefore, the exit strategy for many technology ventures may be an IPO to satisfy the needs of sophisticated investors. Also, it is a way to recapture the management team's time and cash investments. Sometimes, the management team structures the deal so its members receive a portion of their investment from the funds received when the offering is sold.

SELLING THE COMPANY

Selling a company is both a financial and emotional event. The venture team has spent much time, money, and effort in giving birth to the venture. The start-up team has been formed, has attracted necessary resources, executed the business model, hired the staff, and spent precious hours growing the company. Selling the venture is like losing a child. Two primary ways to sell the company are selling the stock and structuring a phased sale.

SELLING THE STOCK

If the company is privately owned (not an IPO), the founder can sell his or her stock to other shareholders at current market rates if the shareholders' agreement allows for this type of transaction. If the company is publicly traded, it's easy to sell the stock on the open market. Two points should be remembered if this exit strategy is used and the company is a public company.

If the founder owns a substantial portion of the issued stock in the company, the strict guidelines from the Securities and Exchange Commission must be followed when he or she decides to sell it. Otherwise there could be stiff penalties.

If the company does have a successful IPO, the founder's stock will have increased significantly in value, which means a serious tax liability. Sometimes, it might be wiser for the founder to sell what he or she needs to meet financial goals and keep the rest of the stock.

Phased sale

The management team may choose to sell the business back to the parent company by structuring a phased sale so that the management team doesn't turn over the reins all at once. For example, the management team may agree to sell the business in two stages. In the first stage, it sells a portion of the business but remains in control of the operations and continues to grow the venture to the point agreed upon in the sale. At that point, the management team will sell the remaining portion to the parent company. Since this technique is complex, attorneys should be used to structure the deal.

MERGING WITH ANOTHER COMPANY

Joining a company with another has some characteristics of a sale along with many characteristics of a partnership. Depending on how the deal is structured, merging a company with another is a way to cash out some of the venture team's investment and stay on in a management capacity or perhaps just sit on the board of directors of the new company.

To execute a successful merger:

- The two companies should have similar cultures; otherwise when they merge, employees and management will most likely have trouble working together. Merging an entrepreneurially run company with a formal bureaucratically run company will result in multiple problems and misunderstandings about how business should be conducted.
- The two companies should be complementary in what they do, to where joining them produces increased value and a company that is better than either one alone. Merging with a company in another industry is often unsatisfactory.
- It is desirable for the two companies to be located close to each other geographically to facilitate management.

- The company with which the merger is accomplished should be in a growing industry and market.

If the company is sold to a public company, usually the public company will want to purchase the company with stock. The company to be purchased is valued at $10 million, and the purchaser's stock value is $10 per share. The company then would receive 1 million shares of stock. The new-venture manager owns 20 percent of the company; therefore, he/she would receive 200,000 shares of the public company and could trade those shares on the public market.

LIQUIDATING THE BUSINESS

Liquidating the business results from a variety of reasons such as the management team may not choose to sell the business because it does not want anyone else to own it. Or, the business may be small in a fragmented industry, so the value is not sufficient to bring a lot of capital beyond the purchase of inventory and equipment. Or the business may be in financial trouble and liquidation through a Chapter 7 bankruptcy filing or reorganization through a Chapter 11 filing may be the only option. Lastly, the management team may find itself in a squeeze-out situation where other stakeholders who have controlling interest in the company are forcing the sale. There may be no available buyers who agree to the terms of sale.

REVERSE MERGERS: LIQUIDATION WHEN THERE'S LITTLE VALUE

Some venture become part of a larger public shell. A shell is a corporation registered with the SEC, but which does no actual business. If the venture team merges with a public shell, it can band with a group that has common goals to eventually gain liquidity by selling stock. After enough small companies are acquired, the shell does a public offering and the small-company owners receive proceeds from the sale of their stock. This is a more complex transaction that requires an experienced SEC attorney.

BANKRUPTCY: LIQUIDATION WHEN THE COMPANY CANNOT SURVIVE

Bankruptcy is certainly the last thing that any business owner wants to face. But the reality is that when the business is upside down and there's no conceivable way to save it, bankruptcy may be the only solution to liquidate the assets of the business and discharge most of the debt.

All the different types of exit strategies help the management team achieve its business goals and objectives. As the business grows, exit strategies and plans change so the management team must continually revise and update them. It would be easy if a business would grow a constant 10 percent every year which would make the value of a business very predictable. Businesses don't grow like this. Instead, they are in continuous flux adjusting to changing economic and marketplace

SUMMARY

- Start developing your exit strategy before or directly after the new venture is launched.
- Exit strategies drive the business from the start-up phase until its exit.
- Review and updated the venture's exit planet least on an annual or semi-annual basis.
- Equity investors require that the founder(s) have an exit strategy in place before investing.
- Companies are not sold at a high value by accident; most of them have developed excellent exit strategies and exit plans.

Chapter 16:

Final Thoughts

Building a life science or healthcare company or commercializing a bioscience technology within your organization has been called "the most difficult relay race in business". As we have noted in the previous chapters, it requires multiples steps that are part of a process that evolves from one step to the next in an extremely risky environment. In addition, few people have all of the knowledge, skills, and abilities to do it all. Consequently, teamwork is essential and team members must include those with different skill sets, whether they are scientific, financial, accounting, marketing, or sales that can help move the invention or project forward.

In this book, we have presented the steps in the Life Science Innovation Roadmap. Embarking on this journey requires perseverance, grit, and courage. The success rate is small, there is usually a long timeline and the landscape is hostile, yet success is extremely rewarding. Hundreds of new biomedical discoveries and inventions make it to market each year, contributing to the US economy as being the global leader in healthcare innovation and the largest market for these products and services.

We have tried to emphasize critical commercializing themes throughout this book and they are worth repeating in this list of fundamental bioscience success drivers.

FUNDAMENTAL BIOSCIENCE SUCCESS DRIVERS

1. The basics of any business are meeting or creating an unmet market need, creating a product or service that offers an attractive value proposition, and using a business model that generates enough profit to sustain and/or grow the business.
2. Planning requires making assumptions about your product/service, the industry, competitors, and customers or end users. Be sure to validate those assumptions as soon as you can and use them as the basis for your planning, financing, and budgeting.
3. Kill your idea early and often, moving on to the next idea if the first one does not work. Often the go/no go decision does not jump off of the page and may not be clean cut. Major roadblocks and hurdles emerge. Don't be discouraged. Rarely does a feasibility plan, for example, come out overwhelmingly positive. It should be used to uncover the Achilles heel of your invention and as a way to gauge whether you will encounter roadblocks and then consider ways to overcome them.
4. Planning is essential. Success, however, depends on finding the right people who can execute. Like football, the best laid plans won't get you to the goal line if your team makes mistakes, gets penalized often or turns the ball over several times.
5. Consider and incorporate your regulatory and reimbursement strategy into your plan as early as possible, not as an afterthought. Getting FDA approval to make and sell your product and getting payment approval are two key value inflection points in the life of bioscience invention or biomedical company. Failure to get these approvals often results in the demise of your company and getting to market.
6. A technology is not necessarily a product and a product is not necessarily a business. Be sure to think through

the best commercialization strategy before proceeding. You most likely won't be able to build and sustain a company around a "one trick pony".

7. While R&D and startup capital is hard to get, there is plenty of money for those with truly disruptive ideas or products.
8. Not everyone is cut out to be a bioentrepreneur. While entrepreneurship can be learned, the most successful have innate talents. Do what you do best and delegate the rest.
9. Life science entrepreneurs are incessant networkers, question assumptions, are good at connecting trends and innovations in industries other than their own, constantly are experimenting and learning from their mistakes and have a knack for creating the future.
10. Plan A almost never works.

Innovation is the life blood of healthcare. Entrepreneurs that create life science companies can rest on the knowledge that they are contributing to the betterment of mankind, stimulating the global economy, creating jobs and, in some instances, leaving a lasting legacy of healing.

We wish you continuing success on that pathway. Good luck with your new venture.

About the Authors

Courtney Price, PhD

Courtney Price, PhD, specializes in management consulting working with physician and administrator leaders to design and implement innovation strategies and systems for healthcare organizations to improve performance, commercialize technologies, and create new sources of revenue. She has been an entrepreneur for the last 25 years, started numerous businesses, and served on corporate boards. In addition, she has been a Professor of Management and the author of numerous several healthcare management and innovation books. She specializes in the fields of innovation, technology commercialization, entrepreneurship and creativity and is a keynote speaker at international, national and regional conferences.

Dr. Price has written more than 10 books on entrepreneurship for John Wiley and Sons, and McGraw-Hill. She was a nationally syndicated columnist of "Entrepreneurs Ask", a weekly column appearing in more than 350 Scripps Howard Newspapers. She founded the Entrepreneurial Education Foundation in 1994 and co-founded Premier FastTrac™ in 1986, the leading business training and development program for entrepreneurs. She became the first Scholar-in-Residence at The Kauffman Foundation and serves as visiting professor at major universities around the world. Price currently is president and CEO of VentureQuest Ltd., a management consulting and training firm that helps organizations introduce new products/

services while increasing, revenues by capitalizing on developing people, technologies and market opportunities.

Arlen Meyers, MD, MBA

Arlen D Meyers, MD, MBA is Professor of Otolaryngology, Dentistry and Engineering at the University of Colorado Denver. He is the Director of the Program in Bioinnovation and Entrepreneurship at the Graduate School at the Anschutz Medical Campus, the founding President and CEO of the Society of Physician Entrepreneurs, a serial bioentrepreneur and consultant, and Director of the Healthcare/Bioscience Division of VentureQuest. Dr. Meyers is a Fulbright Scholar, an alumnus of the Harvard-Macy Program, and was voted by Modern Physician Magazine as one of the 50 most influential physician executives of 2011.

Index

Related Titles from Logos Press®

http://www.logos-press.com

Great Patents

Advanced Strategies for Innovative Growth Companies

David Orange, editor
Softcover: 978-1-934899-18-2
Hardcover: 978-1-934899-17-5

Get to Market Now!

Turn FDA Compliance into a Competitive Edge in the Era of Personalized Medicine

John Avellanet
Softcover: 978-1-934899-12-0

Best Practices in Biotechnology Business Development

Valuation, Licensing, Cash Flow, Pharmacoeconomics, Market Selection, Communication, and Intellectual Property

ISBN: 978-09734676-0-4

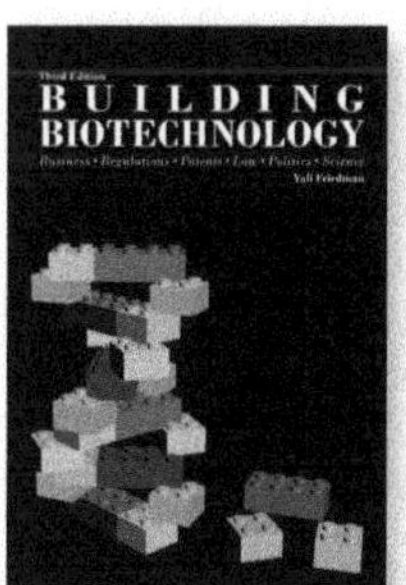

Building Biotechnology

Scientists know science; businesspeople know business. This book explains both.

Hardcover ISBN: 978-09734676-5-9
Softcover ISBN: 978-09734676-6-6

www.ingramcontent.com/pod-product-compliance
Ingram Content Group UK Ltd.
Pitfield, Milton Keynes, MK11 3LW, UK
UKHW020812290926
15470UKWH00003B/1114

9 781934 899267